大兴安岭
野生动物的故事

马敬能（John MacKinnon） 卢和芬◎编著
马千里（Alex MacKinnon）◎绘图

WILDLIFE TALES OF
DAXING'ANLING

科学出版社
北　京

图书在版编目（CIP）数据

大兴安岭野生动物的故事 = Wildlife Tales of Daxing'anling：英汉对照 / 马敬能，卢和芬编著，马千里 绘图. —北京：科学出版社，2018.6
ISBN 978-7-03-057614-9

Ⅰ. ①大… Ⅱ. ①马… ②卢… Ⅲ. ①大兴安岭—野生动物—青少年读物—英、汉 Ⅳ. ① Q958.523.5-49

中国版本图书馆CIP数据核字（2018）第117992号

责任编辑：牛　玲　张翠霞 / 责任校对：邹慧卿
责任印制：赵　博 / 封面设计：有道文化
编辑部电话：010-64035853
E-mail：houjunlin@mail.sciencep.com

科学出版社 出版
北京东黄城根北街 16 号
邮政编码：100717
http://www.sciencep.com
涿州市斑润文化传播有限公司印刷
科学出版社发行　各地新华书店经销
*
2018年6月第 一 版　开本：880×1230　1/16
2022年2月第三次印刷　印张：9 1/4
字数：120 000
定价：98.00元
（如有印装质量问题，我社负责调换）

序 言

　　大兴安岭地处中国的东北，位于黑龙江省西部和内蒙古自治区东北部。这里有着中国面积最大（超过19万平方千米）的连续的森林。大片的落叶松林沿不同方向，弯弯曲曲蜿蜒伸入黑龙江流域，形成一幅犹如马赛克拼接而成的壮丽画卷。

　　大兴安岭的山脉不是高耸如锯齿状的山峰，而是连绵起伏的山峦。森林主要由落叶松组成，间杂着一些白桦和白杨。在平缓一些的山谷里有着蒙古栎树、柳树和榛子树；而地势较高的地方只有一些矮化树，因为在那里一米以下的土地就是永久冻土，树根不能深入土壤。河流和山谷之间分布着沼泽和湿地，干燥一些的土地被开垦成农田。

　　由于地广人稀，生态环境良好，大兴安岭成为动物的乐园。在这片广袤的土地上，繁衍生息着驯鹿、驼鹿、棕熊、水獭、鸮、黑嘴松鸡等各种珍禽异兽，这里也流传着各种关于动物的美丽传说。

　　这本书是专门为儿童创作的，希望能让更多的小朋友熟悉中国北方森林里和湿地中生活着的迷人而美丽的动物。在每个故事的结尾处附有一些简短的问题，小朋友们能够在故事中找到答案。

　　这本书采用了中英文对照的方式，希望小朋友们在学习野生动物保护知识的同时，也能提升中英文的阅读能力。

　　本书是"增强大兴安岭地区保护地网络有效管理项目"的重要成果。我很高兴看到该项目的实施能够加强大兴安岭地区保护地的管理。不过需要强调的是，教育是项目实施的一项重要内容。我有幸担任该项目首席技术顾问，并多次访问了中国大兴安岭这个美丽的地方。我希望这本小小的故事集能传达出我对这片土地的着迷，也能捕捉到大兴安岭当地居民与他们的自然遗产之间正在消失的一些联系。

马敬能

2018 年 4 月

Preface

Daxing'anling is a very special region of China in the extreme northeast. This is China's largest continuous forest covering a landscape of 190,000 sq. km. divided between the two provinces of Nei Menggu and Heilongjiang. The landscape is a mosaic of larch forests draining in different directions into the arched bend of the Heilongjiang river.

The mountains form rolling hills rather than jagged peaks and the forests are mostly comprised of deciduous larch with some birch and poplars. Milder valleys have Mongolian oaks, willows and hazel groves. Higher lands have only stunted trees as roots cannot penetrate deep into the soil which is permanently frozen below 1 metre depth. The rivers valleys form swampy wetlands and drier areas of the landscape have been converted into farmland.

The overall human population in this region remains very small and the landscape is essentially very green and very wild. Daxing'anling has become a paradise for animals. This vast expanse of land is the home for various rare birds and animals, such as red deer, reindeer, moose, elk, deer, and brown bear. Legends about animals and humans are still being told today.

This book is written for children as an introduction to some of the fascinating and beautiful creatures of these northern forests and wetlands. Each chapter ends with a short quiz. Answers to these questions can be found in the text.

This book has Chinese and English texts together. We hope children can improve their reading abilities in both languages, and at the same time, learn the facts about wildlife and conservation.

This book is a product of 'Strengthening the management Effectiveness of the Protected Area Network in the Daxing'anling Landscape'. I have had the pleasure to see the project aims to strengthen the management of protected areas in the region but education is an important component of this objective. I have had the pleasure to serve as technical advisor to this project and made several visits to this beautiful part of China. I hope this small collection of stories can convey some of my own fascination as well as capture some of the disappearing links between Daxing'anling's indigenous people and their natural heritage.

John MacKinnon

April, 2018

目 录

驯鹿怎么来到了大兴安岭
How the Reindeer Came to Daxing'anling ·············· 2

棕熊与鄂温克族人
The Relationship Between Brown Bears and the Evenkis ·············· 16

冰下捕食的水獭
The Otters That Hunt Under the Ice ·············· 26

为什么鄂温克族人惧怕丘鹬
Why the Evenkis Fear the Woodcock ·············· 38

中国驼鹿最后的庇护所
Last Refuge for Chinese Moose ·············· 48

乌林鸮怎样捕食
How the Great Grey Owl Catches Its Prey ·············· 60

狐狸和貂熊
The Fox and the Gulo Gulo ·············· 72

松芽早餐——黑嘴松鸡的故事
Pine Shoots for Breakfast—a Capercaillie Story ·············· 84

傲窦的小狼
Aodu's Little Wolf ·············· 98

天鹅之家
A Hotel for the Great Swans ·············· 114

紫闪蛱蝶奇特的生命周期
The Curious Life Cycle of the Large Blue Butterfly ·············· 124

如何保护大兴安岭美丽的野生动物
How You Can Help Protect the Beautiful Wildlife of Daxing'anling ·············· 138

驯鹿怎么来到了大兴安岭

　　小傲窦从她的帆布帐篷里向外观望，自家养的驯鹿正围着一团闷烧的火堆，躲避讨厌的苍蝇和蚊虫的叮咬。小傲窦看着驯鹿，突然脑子里冒出一个很奇怪的问题。她问道："毛歇叔叔！驯鹿是从哪里来的？"毛歇叔叔温情地看着小傲窦，她已经长大了，是时候跟她讲讲驯鹿的故事了。

How the Reindeer Came to Daxing'anling

As Little Aodu watched from her canvas tent, the small household herd of reindeer huddled around a smoldering fire, flicking away the troublesome flies that tried to bite them. She wondered how it had all begun and suddenly asked her uncle.

"Uncle! Where did the reindeer come from?"

Uncle Mao Xie looked fondly at his niece. She was really growing up so he decided to tell her the full story.

大约4000年前，在中国西北部阿尔泰山地区，驯鹿被北亚人所驯化，或者也可以说是驯鹿驯化了人？事实是，人类从来没有像养羊一样把驯鹿用栅栏围起来驯养。人类想要与驯鹿一起生活，唯一的方法就是跟随它们的足迹在广阔的山脉林地间迁移。因为驯鹿要不断地寻找细嫩且生长缓慢的苔藓，这是驯鹿的首选食物。在夏季，它们还要竭尽全力地寻找到阴凉且没有太多苍蝇、蚊虫的地方。

早先，人们设帐扎营，点燃火堆。驯鹿靠近帐篷和火堆，逃离了苍蝇，也得到了安全——不用再害怕被旷野、森林中凶恶的狼和熊捕食。而人们也发现，有了驯鹿，他们可以轻松地获得很多东西——除了鹿奶，如果偶尔有驯鹿死亡，还可以得到鹿肉、鹿皮、鹿骨和鹿角。此外，人类还发现，驯鹿可以用来驮运货物，甚至运送儿童或拉雪橇。

就这样，驯鹿找到了人类，人类也交上了驯鹿这个好朋友。

在西伯利亚和欧洲广阔的北方森林，农业生产原本是不可能的，人类要想在这里生存极其艰难。而自从有了驯鹿的友情帮助，人类就能在此生存下来。在这样的环境里，与驯鹿共生存才是人类唯一的生存方式。

就这样过了几个世纪，在驯鹿的帮助下，人类的活动范围进一步扩大，西至芬兰，北至西伯利亚，向东北方到达了中国的东北部。随着驯鹿文化的传播，北亚古老的萨满信仰也传播开来——梭罗子、图腾柱、皮鼓、羽毛，以及有神灵和恶魔的宇宙。萨满（巫师，同时也扮演着部落里医药师的角色）据说能够与神灵沟通，能帮助人们作出诸如在哪儿安营扎寨、怎样使人和驯鹿避免疾病和灾难等重大决定。

现在仍然生活在大兴安岭部分地区的鄂温克族人就是这样一个有着古老历史的民族，他们至今仍然拥有驯鹿群，也保留着对远古过去的记忆。

About 4000 years ago, the inhabitants of the Altai Mountains in northwest of China first domesticated the reindeer. Or did the reindeer domesticate man? For indeed humans never trained the gentle reindeer to live in a fenced area. The only way to live with a herd of reindeer is to follow them as they migrate across the wide landscape in permanent search for the delicate and slow growing lichens that form their preferred food and in the summer months in a desperate attempt to find cool shady places without too many biting flies.

By staying near the camp fires of early man, the reindeer escaped the flies and remained safe from the wolves and bears that would prey on them in the wild forests. By tending for the reindeer, people discovered they could get a whole range of useful products—milk, occasional meat, useful skins, leather, bones and horn whenever a reindeer died. Moreover by taming the deer, men found they could use the animals to carry goods, even children or pull snow sleighs.

Reindeer discovered man and man found a good friend in the reindeer.

With the aid of their reindeer friends, man discovered they could now inhabit the vast northern forests of Siberia and Europe where agriculture was not possible and alliance with reindeer was the only way to survive.

And so over the centuries, reindeer helped humans to spread further north and east and west—from Finland in the west to Siberia and northeast of China in the east. And with the reindeer culture spread also the ancient shaman religions of north Asia—the wigwam tents, the totem poles, the leather drums, decorative feathers and belief in a cosmos of spirits and demons. Shamans (Who also plays the role of a tribal pharmacist) could communicate with these spirits and helped the people make big decisions about where to settle and how to avoid diseases of both people and their deer.

One such people are our own ancestors the Evenki that still live in parts of Daxing'anling and still to this day retain herds of reindeer and a memory of their ancient past.

马妮德是小傲窦的奶奶,她已经78岁了。这时她也走过来跟小傲窦和毛歇叔叔一起聊天。她回忆到,在她小时候,中国和俄罗斯之间还没有划分边界,鄂温克族人和他们的驯鹿可以在这片土地上自由来去。现在黑龙江这条巨大的河流把两个国家分隔开了。但到冬天的时候,额尔古纳河(黑龙江的上游)会结上厚厚的冰,驯鹿还是可以轻易地走过去。

Ma Ni De, Aodu's 78 years old grandmother joined into the narrative. She explained that in her childhood there was no border between Russia and China and in those days the Evenki with their reindeer wandered freely between the two countries. Sure there is the great Heilong River that now separates two countries. But the upper reaches such as Erguna river freeze over in winter and the reindeers could simply walk across.

也许在更远古的时候，驯鹿还是大兴安岭的一种野生动物。但到了今天，它们只在鄂温克族人的驯养下生活。偶尔会有一两只驯鹿从鹿群中跑出来，想在野外单独尝试一下自己的运气。

岩石和悬崖上留下来的石器时代的岩画记录了鄂温克族人的祖先用弓和箭狩猎马鹿和野羊的场景。但他们从来不猎捕驯鹿，因为他们认为驯鹿更适合作人类生活的伙伴。

毛歇叔叔继续讲他的故事。

驯鹿有不少与其他种类的鹿不同的地方。首先，它们天性温驯，雄鹿不会为了争夺配偶而大打出手。其次，无论是雄性还是雌性驯鹿都长着大大的鹿角。

Maybe in older times the reindeer was a wild animal in Daxing'anling, but today they remain only in the herds of the Evenkis or as occasional feral escapes that prefer to try their luck alone in the wild.

Whilst stone-age carvings on rocks and cliffs show pictures of the ancestors of Evenki hunting red deer and ibex with bows and arrows, but they never needed to hunt the reindeer. Because they discovered they were more use alive as partners.

Uncle Mao Xie continued his story.

There are several reasons why reindeer are so special. Firstly they are naturally quite tame and gentle. The males do not have great battles over the females and unlike all other deer the females of the reindeer have large antlers almost as big as the males.

驯鹿的食物也很独特，这可能也是导致它们四处迁移的原因。它们生存在极北部，冬天又长又冷，一连几个月都没有鲜嫩的绿叶。在这样的环境下，驯鹿学会了在冬天啃食在积雪下能够长时间保鲜的苔藓。但以苔藓为食有一个非常严重的问题：苔藓生长得极其缓慢，一群驯鹿很快就能吃光一个地方的苔藓，之后它们必须继续寻找新的苔藓生长的地方。

因此，驯鹿的主人必须在照顾驯鹿的同时也关注苔藓的生长。这时候，人能给驯鹿以帮助，因为人会爬树，他们把树上的苔藓采割下来，或把长满苔藓的树枝砍下来好让驯鹿能吃得着。

The food of reindeer is also special and may be the reason that they became so nomadic. Living so far north with long harsh winters, there is no fresh green growth for months on end but the reindeer learned to feed on the leathery lichens that remain fresh under the snow all winter long. The only problem with lichen is that they grow very slowly. A herd of reindeer quickly exhausts the food supply in one area and must move on in search of new un-browsed sites.

Thus tending reindeer became as much a matter of tending the lichen growth as watching after the herds. Man was a good partner too in that he could reach higher up the trees to harvest lichens that were out of reach of the deer, or lop off lichen laden branches for the deer to feed on.

关注苔藓的生长还有另一层意义。苔藓是高度易燃物,所以对于驯鹿和驯鹿的主人来说,没有什么比森林火灾更糟糕的事了。即使是一场小火也足以烧毁一大片驯鹿赖于饲食的苔藓。很久以前,春季干燥,雷电能够引发森林火灾。鄂温克族人总是能发现火情,并很快把火扑灭。而如今的森林比远古那种浓密、潮湿、黑暗的森林更加易燃。因为森林中的树木被砍伐,森林中出现了空隙,风轻易能吹到林下植物和干枯的树枝落叶。现在森林火灾很常见,多是由于人为疏忽造成的,如果有人在森林里不小心遗留了火种,就极易引发森林大火。至今,鄂温克族人都是防范森林火灾的最好的守护者。现代的林业人员可以从鄂温克族人这里学到很多有关大兴安岭森林和湿地的可持续性管理方面的知识。

讲完驯鹿的故事,毛歇叔叔又回帐篷去吸他的烟斗了(噢!这可不是一个好习惯!)。药草和烟草的味道混杂在一起弥漫开,宁静的氛围充满了帐蓬。

Tending for lichens had another implication. Lichens are highly flammable, so nothing was worse for the deer and human masters than a forest fire. Even a small fire can destroy a wide area of lichen that the reindeers live on. There were forest fires since ancient times caused by lightening striking in the drier spring season. But always the Evenki were quick to spot the fires and put them out. The forests today are easier to burn than the tall dark damp forests of former times, because the forests are now logged and opened up, allowing winds to blow through the undergrowth and dry the fallen branches and leaf litter. Today fires are common and often caused by human carelessness. Any source of fire accidentally left in the forest could easily trigger a forest fire. To this day the Evenkis are the best protectors of the forests against the spread of fire. There is much the modern foresters could learn from the Evenkis about sustainable management of the forest landscape and wetlands of Daxing'anling..

Uncle Mao Xie went back to his pipe after his story. (Oh! This is not a good habit!) The sweet scent of the fragrant leaves that he mixed with tobacco filled the tent with a peaceful reassurance.

小测试

1. 是谁最早驯化了驯鹿？

2. 驯鹿如何能跨越黑龙江？

3. 驯鹿喜欢吃的食物是什么？

4. 为什么驯鹿老围在火堆旁边？

5. 你能列出四样人们可以从驯鹿身上获得的东西吗？

6. 为什么鄂温克族人住在帐篷里？

7. 萨满在鄂温克族人中的作用是什么？

8. 为什么驯鹿的主人很注重保护森林避免火灾？

答案

1. 大约4000年前居住在中国西北部阿尔泰地区的北亚人；

2. 在冬季，黑龙江的上游会结冰，驯鹿可以轻易走过去；

3. 它们吃在森林地面和低树枝上生长的苔藓；

4. 可以减少苍蝇和蚊虫的叮咬；

5. 鹿角、鹿奶、鹿皮、鹿肉；

6. 因为他们必须跟随驯鹿群不停地迁移，住在便于携带的帐篷里是唯一的方法；

7. 巫师，同时也有医药师的作用，据说他们能连接神灵世界和凡人世界；

8. 因为即使是一场小火也会摧毁一大片生长缓慢但驯鹿又赖于饲食的苔藓。

Quizz Time

1. Who first domesticated reindeer?

2. How can the deer cross the Heilongjiang river?

3. What is the preferred food of reindeer?

4. Why do the deer sit around fires?

5. Can you list four products that people can obtain from reindeer?

6. Why do the Evenki live in tents?

7. What is the role of shaman in Evenki society?

8. Why are reindeer herdsmen keen to protect the forest from fire?

Answers

1. North Asian people who lived in the Altai region of northwest of China about 4000 years ago;

2. In winter the upper branches of the Heilongjiang river freeze over making it easy for reindeer to walk across ;

3. They eat the lichens that grow on the forest floor and low tree branches;

4. They have learned that smoke limits the biting flies and mosquitos;

5. Horn, milk, hides, meat;

6. They have to keep moving from place to place following their herds of reindeer, so portable tent is the only way;

7. They act as medicine man and act as a go between linking the spirit world and human world;

8. Even a small fire destroys the slow-growing lichens on which reindeer feed.

棕熊与鄂温克族人

对于鄂温克族人和其他生活在亚洲东北部的古老民族来说，森林里最令人敬畏的动物不是狼而是棕熊。棕熊就是动物之王。鄂温克族人认为：在远古时期，除了人类，棕熊是森林中唯一能够思考和说话的动物，因此人与棕熊可能拥有共同的起源。同许多其他民族一样，鄂温克族人认为北方星系是熊的星系。在那片土地上，棕熊为他们引路，在夜晚给他们指引方向。

The Relationship Between Brown Bears and the Evenkis

To the Evenkis and other ancient peoples of northeast of Asia, the most frightening animal in the forest was not the wolf but the brown bear. The brown bear was the king of the animals. The Evenkis believed bear was the only other animal in the forest able to think and talk and that man and bear share many ancestral relationships from the earliest origins of living things. Like other peoples, they regard the northern star cluster as the constellation of the bear. So it is the bear that guides them across the landscape and gives them their sense of direction at night.

棕熊是这片森林中最大的食肉动物，站立起来有 2 米多高，重量可达 500 千克。雄性棕熊的体重是雌性的两倍。雌性棕熊从 5 岁开始繁殖，雄性从 6 岁开始。棕熊的寿命较长，在野外平均寿命可达 25 年。

棕熊常常独处，但活动范围很大。一只棕熊可能需要 300 平方千米的领地来活动。它们只在秋季的几天里交配。雄性棕熊不参与抚养幼畜。

Brown bears are the largest carnivores in these forests and can stand more than 2 metres tall and weigh up to 500 kg. Male bears are twice the weight of females. Females breed at 5 years; males usually at 6 and both sexes are long lived with wild animals averaging about 25 years.

Brown bears are rather solitary and roam over huge ranges. One bear may need a home range of 300 square km. Their mating takes place for only a few days each autumn. The male plays no further role in bringing up the two to four resulting cubs.

棕熊并不冬眠。在冬季的几个月里，它们爬入洞穴、空心树洞或其他庇护地慵懒地生活，这时候，它们的能量供应基本来自在夏季和秋季储存积累的脂肪。第二年的春天来临时，它们的体重通常只剩下原来的一半。

正在养育幼崽的雌性棕熊攻击性很强，常常为了保护幼熊而能驱逐体形比她大得多的雄性。

棕熊吃各种各样的食物，包括浆果、蘑菇、蠕虫，以及它们能捕捉或挖掘到的动物，它们也会啃食动物的尸体。偶尔，为了捕食，棕熊也会对温驯的驯鹿发起攻击。

几十年前，那时候法律还没有禁止捕杀棕熊，但出于对棕熊伟大智慧的尊重和它们与人类祖先的关系，鄂温克族人通常不会捕杀棕熊。如果在森林里遇到一只幼熊，他们会收养它，给它穿上衣服，把它作为家庭中的一员来抚养，直到它长到足够大才会带到野外去放生。如果有棕熊混入驯鹿群，要驱赶棕熊离开还得由萨满主持仪式。

如果真的迫不得已要杀死一只棕熊，鄂温克族人也会举行仪式来请求棕熊的宽恕并保证棕熊死后能够重生。棕熊的头骨必须挂在朝东向阳的树上，鄂温克族人认为这样太阳就能将其灵魂托付到一个新的躯体上。一只熊被杀后，鄂温克族人会接连几天举行繁复的活动，既安抚熊的灵魂，又享受丰盛的肉宴。

鄂温克族人的驯鹿群很小，它们聚集在营火周围，以远离它们的天敌。此外，鄂温克族人也养育大型犬，晚上若有入侵者，它们会吠叫并发起攻击。但即使这样，冲突还是不可避免。马妮德老奶奶告诉小傲窭，曾经有一只棕熊就不愿远离他们，常常来捕食她家的驯鹿。这使得家里驯鹿的数量急剧下降。为了驯鹿和家人的生存，他们最终不得不决定杀掉这只棕熊。

Brown bears do not hibernate as such but enter into caves, hollow trees or other shelters to become lethargic through the winter months living off fat they have stored in the summer and autumn and losing up to half their weight before the next spring arrives again.

Females with cubs are very aggressive and will even attack and chase away much larger males in protection of their young.

Brown bear eats a wide range of foods including fruits, mushrooms, worms, animals it can catch or dig up and they will also scavenge on the remains of dead animals. Occasionally a bear becomes a real predator and finds the tame herds of reindeer an easy prey.

A few decades ago, there was no law banning the killing of brown bears. But, out of respect for the bear's great intellect and relationships with the mother creator, Evenkis usually did not kill bears. If they found a young bear in the forest they would adopt it, dress it in clothes and bring it up as a member of the family until it was large enough to release in the wild. The shamans would officiate over ceremonies to drive bears away from their deer herds.

If a bear really had to be killed then there were other ceremonies to seek forgiveness for this act and to ensure that the bear would be reborn. The skull of the bear had to be hung on an east facing tree so that the sun could regenerate its spirit into a new body. When a bear was killed, a complicated celebration lasting several days ensued, both to placate the bear's spirit and to enjoy the rich bounty of meat.

The Evenki herds are small and gather around a camp fire which keeps most predators at bay. In addition the Evenkis keep large dogs that will bark at and attack intruders at night. But conflicts do occur and old Ma Ni De told little. Audo a tale of one bear that simply would not stay away and kept coming back to take more and more of the family's fast dwindling deer herd. The only way for her and their reindeers to survive was to kill the bear.

这事由她的表弟贺协负责解决。在举行了适当的仪式后,他们准备要杀死这只熊。贺协扛着枪,马妮德老奶奶带上手电筒。表弟很肯定棕熊会在午夜时分出来。他们爬上一根粗壮的树枝,在黑暗中等待,马妮德老奶奶害怕极了。

正如贺协所料,刚到午夜,他们就听到一只大型的动物在朝他们靠近。"打开手电筒!"她的表弟低声说,马妮德老奶奶打开手电筒,灯光一阵狂摆乱晃后,集聚到了这只有粗毛的棕色动物身上。

棕熊转过身来直接面对他们。贺协开了一枪,但没有击中它。棕熊迅速地跑开了。从此,这只棕熊再也没有回来过。马妮德老奶奶的鹿群安全了,很多年再没有棕熊来骚扰。

今天,鄂温克族人不再允许用枪狩猎。棕熊在大兴安岭地区也相当罕见了。

Her cousin He Xie took charge of the operation and after appropriate ceremony made preparations to kill the bear. Cousin carried a gun and Ma Ni De carried a strong torch. The cousin He Xie was sure the bear would come at midnight. Ma Ni De was very frightened as they waited in the darkness perched on a stout tree branch.

At midnight, just as He Xie had predicted, they heard a large animal approaching. "Now" whispered her cousin. Ma Ni De switched on the torch swinging the beam wildly before focusing on the shaggy brown animal.

It turned to face them. He Xie fired one shot, but missed it. The bear ran away quickly. From then on, the brown bear never came back again. Ma Ni De's herd has been safe from bears for many years now.

Today the Evenkis are no longer allowed to hunt with guns but today the brown bear is rather rare.

小测试

1. 狼和熊，鄂温克族人更尊重哪一个？

2. 鄂温克族人认为天上哪个星系是熊的星系？

3. 雄熊会帮助养小熊吗？

4. 棕熊在冬天会冬眠吗？

5. 为什么棕熊雌熊体形不大，却敢跟体形大她很多的雄熊打架？

6. 棕熊吃什么？

7. 为什么马妮德老奶奶他们想要杀掉熊？

8. 为什么鄂温克族人需要很大的狗？

答案

1. 他们厌恶并且害怕狼，但是他们尊重熊，因为人与熊有共同的起源；

2. 北方星系；

3. 不，雌熊独自养育小熊。

4. 不是真的冬眠，而是变得慵懒，不活跃；

5. 保护她的幼熊，因为一只大的雄熊可能会杀死并吃掉幼熊；

6. 熊吃各种各样的食物，包括浆果、蘑菇、蠕虫，以及它们能捕捉或挖掘到的动物，它们也会啃食动物的尸体；

7. 因为熊不断回来袭击她家的驯鹿群；

8. 大的狗会对入侵的大型食肉动物如熊和狼等发出警告。

Quizz Time

1. Which do the Evenkis most respect, wolf or bear?

2. Where in the sky can you see the bear constellation?

3. Does the male bear help to raise the cubs?

4. Do bears hibernate in winter?

5. Why would the small female bear fight the bigger male bear?

6. What do bears eat?

7. Why did Ma Ni De help kill a bear?

8. Why do the Evenkis need big dogs?

Answers

1. They hate and fear wolves but they respect the bear as a spiritual relative;

2. The extreme north;

3. No, the female bear raises the cubs alone ;

4. Not really hibernate but sluggish and inactive;

5. To protect her cubs, a big male might kill and eat them;

6. A wild range of foods including fruit, mushrooms, worms, animals it can catch or dig up and they will also scavenge on the remains of dead animals;

7. Because it was coming back to raid the family reindeer herd;

8. The big dogs warn of enemies and big carnivores such as bears and wolves.

冰下捕食的水獭

一个夏天的晚上，奥利和他的父亲健科坐在一根圆木上，看着缓缓流淌的河水。奥利看见了一种罕见的动物——一只母水獭和三只小水獭沿着河边游泳玩耍。

"蹲下，安静，水獭非常怕生，非常敏锐，我们就看着它们玩吧。"爸爸轻声对奥利说。

父子俩瞪大了眼睛静静地盯着看了一会儿，直到他们听到一声尖哨声，水獭一家急急忙忙下了河滩，迅速溜到河道拐弯处。

那天晚上，在帐篷里营火旁，奥利问了许多关于水獭的问题，他的父亲细细地给他讲了有关这些可爱的动物的知识。

The Otters That Hunt Under the Ice

One summer evening as Aoli and his father Jian Ke sat on a log overlooking the slowly swirling river, Aoli caught sight of a rare animal—a mother and three young otter cubs swimming and playing along the river's edge.

"Crouch down, keep quiet. Otters are very shy and very sharp sensed. We can watch them play." Dad whispered to Aoli.

The two pairs of human eyes watched silently for a few minutes till with a shrill whistle the entire family scampered down the beach and round the bend of the river.

That evening, by the fire in their tent, Aoli asked many questions about otters and his father gave a fuller explanation about these cute animals.

水獭是一种哺乳动物，皮毛乌黑亮滑，耳朵短，尾巴粗壮。它们跑起来有点儿笨拙，不过它们可是游泳的高手。在水里，它们会使用强健的尾巴推水，用有蹼的脚划动加速，随着水流上下起伏，泳姿优雅，速度非凡。

在岸上，它们的视力并不那么好，因为它们的眼睛适应了在水下看鱼。不过，它们的味觉和听力都足够敏锐。

Otters are sleek black mammals with short ears and a stout tail. They run with an undulating gait but are great swimmers. They use their strong tails for propulsion but also undulate through the water with grace and speed and paddle with their webbed feet for additional speed.

On land their eyesight is not so good because their eyes are really adapted for seeing fish under the water, but their smell and hearing are sharp enough.

"那么水獭是不是只在河边生存？"奥利问道。

"是的，"爸爸回答道："水獭总是在离水不远的地方居住，因为它们要沿着河岸游很长的距离找鱼吃。每家水獭大概需要30千米长的河道。它们会在河岸或树根下面挖掘出好几个洞穴，这些洞在白天可以用来躲藏，到了冬天可以用来越冬。这些洞穴会有一个水下的入口，水獭由此进出。每年水獭都会选出一个洞穴作为繁殖点，水獭妈妈在里面生育她的小宝宝。"

"So do the otters only live by rivers?" asked Aoli.

"Yes. They are never far from water. They roam along many kilometers of river banks searching for fish. A family of otters needs around 30 kilometers of river. They make a series of dens—holes in the river banks or under tree roots where they can hide up in the day or rest in the winter. The caves would have an underwater entrance, and the otters would come in and out. Each year one of these dens will be selected as the breeding den and there the female otter will have her pups." answered Daddy.

幼小的水獭不能独立生活，但它们成长迅速，很快就能游到洞穴外来探索夏日的河岸。虽然还需要依靠妈妈给它们带来食物，但它们喜欢在水中追逐，并沿着河岸游来游去，像淘气的孩子一样。

水獭能捕捉到很大的鱼，但现在这样的鱼已经很少，它们大多只能依靠较小的鱼和螃蟹、小虾和昆虫来填饱肚子。它们常常在突出的岩石下如厕，这些岩石是它们的领土标记，以警告其他的水獭：这段河流已经有主人啦！

"水獭的粪便中有鱼鳞和甲虫的鞘翅，你可以通过这个方法来辨识哪些岩石是水獭的'茅坑'。"

"冬天结冰了的时候它们怎么办？"小男孩认真地问道。

冬天一到，雪就纷纷飘落，河流冻结成冰，但仍然可以在结冰的河床上看到水獭夜里在水道上溜跶时留下的痕迹。河流冻结时，水獭是怎么觅食的呢？那是它们的秘密！

你有没有见过渔民冬天在冰冻的河里捕鱼？渔民们在大的河道中间开一个洞口，冰下的河水仍然在欢快地流淌，渔民们放下饵线或渔网捕鱼。事实上，几乎所有的河流都只表层冻结，尽管在大兴安岭这么寒冷的地方，有的冰层的厚度能超过一米，但冰层下仍然有一小段一小段流动的河水。

此外，随着冬天的水位下降，冰层和下面的流水之间形成一个空隙。水獭因此有充足的空气可以呼吸，也就可以在河流冰层下由隧道和小水池组成的秘密世界中狩猎。

事实上，在冬季，河里的鱼都集中在冰层下的这些未冻结的一个个水池中求生存，这样水獭就更容易找到足够的食物。

"那我们为什么极少见到它们？"奥利问道。

The pups are helpless at first but grow quickly and soon emerge to explore the summer outdoors. They depend on mum to bring food but they like to play chasing in the water and sliding down the river banks like naughty children.

Otters can catch large fish but such fish are scarce now and they mostly have to rely on smaller fish and crabs, crayfish and insects to make do. They choose prominent rocks as their regular toilet and this also serves as a territorial marker warning other otters that this stretch of river is occupied.

"You can find these toilets and recognize the dung of otter by the numbers of fish scales and beetle elytra in them."

"What do they do in winter when the rivers freeze over?" asked the small boy earnestly.

In winter the snow falls and the rivers freeze but still you can see the tracks where the otters have scampered along the riverbeds at night. But how do they find food when the rivers are frozen? That is their secret!

Have you seen the fishermen fishing in the frozen rivers in winter. They cut holes in the ice and there is still flowing water deep underneath. Indeed almost all rivers only freeze on the surface and even though the ice may be as deep as one metre there is still a small flow of water underneath.

Moreover as water level falls in the winter, there is a gap of air between the ice surface and the flowing water underneath. Plenty of air for the otter to breathe and so the otter can hunt underneath the ice in a secret world of tunnels and river pools.

Indeed it is in these unfrozen pools beneath the ice that the river fish concentrate to survive the winter and this makes it easier for the otters to find enough food.

"Why do we see them so rarely?" asked Aoli.

爸爸接着给奥利解释为什么水獭变得这么稀少了。

水獭的毛皮光滑柔软、温暖蓬松，让水獭既可以在冬季完美保暖，又能在游泳时减小水的阻力。因此，水獭的皮毛对人类来说非常珍贵。几个世纪以来，鄂温克族人一直把水獭的皮缝在木制雪橇的底部来减少摩擦，让雪橇能跑得更快；近年来人们对皮草的喜爱使得水獭皮价格高昂，水獭因此被大量猎杀；渔民也因为不想与它们分享河流中的鱼而捕杀它们。

水獭因此变得非常稀少。但水獭真的很可爱！而且它们是这个大自然不可缺少的一员！我们现在要保护水獭，帮助恢复它们的数量。

但是，这并不容易，因为河流里中它们赖以为食的鱼也越来越少了。人们使用网眼很小的渔网和非法电动捕鱼装置，过度捕捞鱼类。加上城镇矿山排放出的污水污物，农田里使用过多的化肥，林场使用过多的杀虫剂，等等，使河流受到严重污染，很多鱼类也因此而丧生。此外，沿着水道建造的许多小型水坝、水堰，阻止了成鱼顺利地返回产卵地繁殖，这也导致了鱼的数量急剧下降。

我们需要先恢复鱼的数量，水獭也就会变得更常见。

奥利突然有了一个好主意："也许我们需要修建养鱼场，这样既能满足人们对新鲜鱼的需求，同时还可以每年向河流放养足够量的鱼，以保护水獭。"

"我们的确应该这么做。"他父亲赞扬他："而且，我们还可以开发一种新兴的生态旅游——休闲性钓鱼来帮助地方经济，但是大部分钓上来的鱼都应该放回到河里去。"

奥利的父亲健科是位于大兴安岭的根河源国家湿地公园的公园监管员。根河源国家湿地公园现已开展更进一步的工作：他们把水獭设为他们保护的主要目标，并通过漫画和可爱的毛绒玩具帮助小朋友们认识和学会保护水獭。

Jian Ke went on to explain why the otter had become so rare.

"The fur of the otter is smooth and silky yet warm and fluffy—perfect to keep the otter warm in the winter but offering minimum resistance to water when swimming. This makes otter skins very precious to humans. For centuries the Evenkis used otter skins to line the underside of their wooden skies for use in winter snow. More recently the demand for animal furs put a high price on the skins of otters. So otters were hunted and trapped for their fur and otters were hunted by fishermen who did not want to share the fish with the wildlife.

Thus otters became very rare. And yet they are really so cute. They are an indispensable member of the nature! So now we need to conserve otters and help them build back their population.

But this is not so easy as the rivers have lost most of their fish. The rivers are overfished by human nets and fished by humans illegally, using electric fishing devices. The rivers are polluted by discharge from towns, mines and too much chemicals on the farmlands or insecticides on forest plantations. Moreover the many small dams and water diversions built along waterways prevent the free flow of fish between their cool upstream breeding streams and the main rivers where the larger adult fish prefer to live.

"We need to get the fish stocks back first and then the otters can again become more common."

Aoli had a great idea. "Maybe we need to set up fish farms so we can meet the needs of local people for fresh fish but can also release enough fish each year into the rivers to allow conservation of otters."

"Exactly what we should do." commended his father. "And then we can also develop a new industry of recreation fishing as a form of eco-tourism to help the local economy but return most of the caught fish back into the river."

Genheyuan Wetland Park where Aoli's father is a park warden has gone even further and now makes the otter a main target for conservation and children learn to love the otter through its cartoon characters and loveable soft toys.

小测试

1. 水獭吃什么？

2. 水獭住哪里？

3. 一个水獭家庭需要多长的河道？

4. 水獭冬眠吗？

5. 冬天鱼类去哪儿了？

6. 渔民们冬天怎么捕鱼？

7. 为什么人们愿出高价购买水獭皮毛？

8. 为什么现在河里的鱼这么少了？

答案

1. 鱼类、蟹类、软体动物和无脊椎动物；

2. 在河边的地下洞穴中，通常有一个水下入口；

3. 每个水獭家需要约30千米河道；

4. 水獭不冬眠，冬天它们更容易在河里捉到鱼；

5. 它们聚集在冰下的没有结冰的河段内；

6. 渔民们在大的河道中间开一个洞口，放下饵线或渔网捕鱼；

7. 水獭皮毛漂亮、温暖、防水好，因为阻力小，可以用来增进滑雪板的润滑；

8. 过度捕捞，河流被污染，筑建水坝、水堰导致河流被截断，鱼无法正常繁殖。

Quizz Time

1. What do otters eat?

2. Where do otters live?

3. How many kilometers of river does each otter family need?

4. Do otters hibernate in winter?

5. Where do the fish go in winter?

6. How do fishermen catch fish in winter?

7. Why is otter fur so prized by people?

8. Why are there so few fish in the rivers today?

Answers

1. Fish, crabs, molluscs and invertebrates;

2. In underground burrows beside rivers, often with an underwater entrance;

3. About 30 km per family;

4. No, they can catch plenty of fish all winter;

5. They get trapped in unfrozen pools under the ice;

6. They cut holes in the ice in the middle of large rivers and drop down baited lines or nets;

7. Otter fur is smart, warm, waterproof and low resistance so it can be used to improve the sliding of wooden skis;

8. The rivers have become polluted, overfished and fragmented by dams and weirs.

为什么鄂温克族人惧怕丘鹬

一天晚上,马妮德老奶奶讲述了一些关于鄂温克族人的信仰和迷信的故事。她抱歉地说她现在老了,忘了很多事情。她说话缓慢,声音沙哑。但是,当她回忆起温馨的往事时,她的双眼闪烁着欢乐。

鄂温克族人认为,整个世界是一个活生生的实体——岩石、土壤、树木、河流和动物都是相互关联的,必须保持平衡。所有的生命都有灵性和魂魄,而魂魄也有善有恶,会根据人们的不同行为给他们以帮助或惩罚。

有一种鸟,看起来不起眼的丘鹬,却被鄂温克族人认为是邪恶魂魄的载体。

丘鹬是一种水禽,体形大而笨重,腿短,嘴长呈锥形;背部为红棕色,腹部为淡黄色。丘鹬的双眼在头两侧较后位置,因而它有着360度视野。丘鹬长得比较像沙锥、矶鹬或塍鹬等其他水禽,但这些种类的鸟很普通,鄂温克族人并不惧怕。

Why the Evenkis Fear the Woodcock

One evening, grandma Ma Ni De told us about some of the Evenki beliefs and superstitions. She apologized that she is old now and forgets many things. She talks with a slow rather cracked voice. But her eyes twinkle with joyful memories as she warms to her subject.

The Evenkis believe that the entire world is a living entity—the rocks, soil, trees, rivers and animals are all connected and must remain in balance. All living things have spirit souls but there are both good and evil spirits that can help or persecute humans depending on their deeds.

One bird especially seems to be the host of evil spirits—the modest woodcock.

The woodcock is a large bulky wader with short legs, and a very long straight tapering bill. Its eyes are set far back on its head to give it 360-degree vision. The woodcock is a wader and looks like many other related waders such as snipes, sandpipers or godwits, but these are common birds that hold no fear for the Evenkis.

丘鹬习性怪异，与众不同。它不像其他水禽栖息在水边，而是生活在林地中；它也不像其他水禽在白天活动，它是夜行性鸟类。白天，丘鹬就深深地隐藏在林地的落叶和杂草中。丘鹬上体为红棕色，下体浅黄，这种颜色形成了它隐蔽的伪装色。体羽上精致的图案也能让丘鹬完美地融入周围的环境，不易被发现。当人们走进林地时，根本看不见丘鹬。只有当人走近了，丘鹬仿佛是从哪里突然冒出来的，猛地腾空飞起，把人吓一大跳。到了晚上，当其他水禽休息了，丘鹬才从林地里出来，飞到肥美的湿地或湿草地中寻找食物。这样，它们就不用与其他鸟类争抢食物，也不用理会其他更活跃的水禽们的聒噪。

它们在黄昏时候一边沿着溪流或森林飞行，一边发出叫声。丘鹬的叫声也很怪异，像青蛙似的呱呱三四次大叫之后，接着发出一声刺耳的尖叫。丘鹬在清晨和傍晚比较活跃，白天极少活动，除非受到惊吓，它就会突然起飞，飞翼发出呼呼叫声。它飞行时与猫头鹰、蝙蝠有些类似，在穿越开阔的田野时，丘鹬飞得又快又直，但在丛林中，它的飞行或扭曲或震颤，极不规范。

The woodcock is special because it is strange. Unlike other waders it lives in forests and unlike other waders it is nocturnal. During the daytime the woodcock sits tight, hiding in the forest undergrowth. It has cryptic camouflage to suit its woodland habitat, with reddish-brown upperparts and buff-coloured underparts. Its intricately patterned plumage blends invisibly into the leaves and grasses. So you really never see the woodcock on the ground and if you do get too close it flies suddenly upwards in startling fashion as though it has come from nowhere. Only when the other waders retire to their roosting sites for the night, does the woodcock come out to fly over the forest and look for nice wetlands or moist grassy fields where it can look for food without the competition and bustle of the more active wader relatives. Woodcocks often call as they fly along the stream—sides or forest rides at dusk. The call too is unusual being three to four deep froglike growls followed by a shrill squeal.

Woodcocks are crepuscular (most active at dawn and dusk) and rarely active during the day unless flushed, when they fly off with a whirring wing noise. The flight is somewhat owl-or bat-like. They fly fast and directly when crossing open country, but fly erratically with twisting and fluttering once in woodland.

丘鹬通常独自生活。在4月和6月之间的黄昏，雄性丘鹬会进行一场求偶飞行仪式。它们一边飞行一边鸣叫。

丘鹬的巢一般在丛林的低处，用枯叶和树枝等筑成，形状像个杯子。雌鸟一窝产4枚蛋，蛋壳是白色或是奶白色，上面有浅棕色或灰色的斑点。鸟蛋的孵化期一般为三周，由雌鸟承担孵化工作。很快，雏鸟就出生了。虽然10天后雏鸟就能短距离地飞行，但要在15～20天之后，它们的羽毛才能丰满。如果感觉到了危险，雌鸟会用爪子抓住雏鸟，把它放在双腿、身体和尾巴之间，或是把雏鸟背在背上一起飞离。这也是鄂温克族人认为丘鹬奇怪的地方。

丘鹬在灌木下松软的土壤里觅食，通常很隐蔽。它们把长长的锥形嘴伸进软土层，去寻找甲虫、蜘蛛、毛虫、蝇幼虫、小型淡水软体动物以及一些植物种子。因为它们在土里觅食，所以当冬季天气很冷，地面被冻得严严实实的时候，它们就很难找到食物。所以每到冬天，它们就会迁徙到遥远的南方去寻找没有霜冻的森林。

也许正因为丘鹬稀少罕见，或是因为它的行为和叫声都非常怪异，鄂温克族人就怀疑它是不是有邪恶魂魄附体了。

They are usually solitary. The male performs a courtship display flight called 'roding' at dusk between April and June. A mating call is performed during courtship display flights: "orr, orr, orr, pist".

Woodcocks nest on the ground in low cover in woodland . The nest is a lined cup or a slight hollow lined with dead leaves and other plant material. A single brood of four white or creamy eggs with light brown and grey blotches is laid; once the clutch is complete, incubation is performed by the female for between three weeks. The downy, precocious young leave the nest immediately and fledge after another two to three weeks, although they can fly short distances after 10 days. When threatened, the mother bird can fly whilst carrying small chicks between her legs, body and tail, in her claws or even on her back. Just another strange habit to convince the Evenkis this is no ordinary bird.

Woodcocks forage in soft soil in thickets, usually well hidden from sight. They mainly eat earthworms, but also beetles, spiders, caterpillars, fly larvae and small freshwater mollusks and some plant seeds. Because they rely on probing into the ground to find food, they are vulnerable to cold winter weather when the ground remains frozen. So at such times the birds migrate further south in search of unfrozen forests.

Perhaps it is so rare to see a woodcock and because its behavior and calls are rather peculiar, but the bird is treated with great suspicion by the animistic Evenki and viewed as a clear candidate for harboring evil spirits.

马妮德老奶奶说,在晚上看到或听到丘鹬并不会让人觉得怪异,但如果在大白天看到或听到这鸟就太奇怪了,对于鄂温克族人来说,这就意味着今天会有厄运。所以一旦看到丘鹬,最保险的就是赶紧回帐篷去睡过这一天。

晚上,鬼鸮的怪异叫声也被视为厄运的预兆,预示着有人或家畜要死亡。有时候晚上与其整夜醒着听着森林里的动静,不如喝了酒后酩酊大睡。

一想到喝酒,马妮德老奶奶就笑了,停下不语。但突然又想起什么,便接着自个儿讲下去。

她想起过去还没有像样的医疗服务时,很多类似的预兆和迹象、或各家有生老病死都会请求族中萨满来处理。萨满有时候是男人,但更通常是女人。鄂温克族人相信:只有萨满才能随烟雾从帐篷顶洞出去,飘到人之上的世界上去找善良的魂魄或到下层世界去找险恶的魂魄,跟他们协商解决问题;也只有萨满能够把死去的人的魂魄送到中间世界去开始下一段生活。

鄂温克族萨满祭司时穿着长袍,上面点缀着色彩浓艳的条条带带。他们不停吟唱,手舞足蹈,同时敲击着一个小圆鼓,或摇晃着金属响环。直至今天,萨满在与神灵世界沟通的秘密祭司时,仍然使用有魔力的咒语和魔鼓。

鬼鸮

Ma Ni De explains that to see or hear the bird at night is normal and accepted, but to see or hear the bird in the daylight hours is strange and considered by the Evenkis to be the signal of impending bad luck. Seeing a woodcock by day and it is safest to hurry back to your camp and sleep for a day.

At night the eerie calls of the Funereal Owl are also regarded as a bad omen—a sign that a person or one of domestic animals will get sick or even die. Sometimes it is better to drink wine and sleep soundly rather than stay awake through the night listening to the sounds of the forest.

At the thought of liquor, Ma Ni De smiles and pauses. Then spurred by other memories, she continues her slow monologue.

She remembers the olden days when there was no formal medical support. Too many signs of evil spirits, or death or sickness in the family all call for the work of the shaman or clan spirit magician. The Shaman is sometimes a man but more usually a woman and it is believed that only the shaman can travel with the smoke through the tent hole to visit the good spirits of the upper world or the bad spirits of the lower world to settle human's affairs. It is the shaman who can escort the spirits of dead people to the next life in the middle world.

The Evenki shaman would wear dazzling clothes of long coloured strands and would dance and sing, whilst beating on a small round drum and shaking metal rattles. To this day shamans still keep decorated charms and a magic drum for use in their secret ceremonies to communicate with the spirit world.

小测试

1. 丘鹬在哪儿找吃的？

2. 丘鹬在树上筑巢吗？

3. 丘鹬冬天到哪儿去越冬？

4. 丘鹬喜欢群居吗？

5. 鄂温克族人认为其他什么鸟的叫声也是厄运的预兆？

6. 为什么人们很少能见到丘鹬？

7. 如果人生病了，鄂温克族人会请谁来看病？

8. 丘鹬怎么能看到身后发生的事？

答案

1. 它在潮湿的田地里或离水近的地方，把长长的锥形嘴伸进软土层，寻找甲虫、蜘蛛、毛虫、蝇幼虫、小型淡水软体动物以及一些植物种子；

2. 不，它们在林间的低处筑巢；

3. 它们向南迁徙到不冻结的森林里越冬；

4. 不，丘鹬喜欢独居；

5. 鬼鸮；

6. 因为它们混在林下的叶子中间，而且只在夜间活动；

7. 萨满；

8. 丘鹬的眼睛长在头部的两侧，因而可以看到360度。

Quizz Time

1. Where does the woodcock find its food?

2. Does the woodcock nest in trees?

3. Where does the woodcock go in winter ?

4. Does the woodcock live in large flocks?

5. What other bird call is taken as a bad omen?

6. Why do you rarely see the woodcock?

7. Who did the Evankis call if someone is sick?

8. How can the woodcock see what happens behind its back?

Answers

1. It probes its long bill deep in soft earth in wet fields or near water to find earthworms, but also beetles, spiders, caterpillars, fly larvae and small freshwater mollusks and some plant seeds;

2. No, they nest on the ground;

3. They migrate south to unfrozen forests;

4. No, woodcock is solitary;

5. The Funereal Owl;

6. Because they are camouflaged and nocturnal;

7. The shaman;

8. Its eyes are placed on the sides of its head and it can see all 360 degrees.

中国驼鹿最后的庇护所

健科是根河源国家湿地公园的公园监管员,他对森林和湿地中的动物了如指掌。他最关注的就是中国驼鹿。他对驼鹿的所有习性如数家珍。驼鹿是他最爱谈论的话题了。

驼鹿是鹿群中体型最大的成员,也是中国最大的陆地哺乳动物之一,分布在大兴安岭和小兴安岭地区,或许阿尔泰山也有。成年驼鹿能达2米高。它们身体硕大,全身有深棕褐色的短毛。驼鹿耳朵很大,脸呈长方形。驼鹿的喉部下面有一个肉柱,称为颔囊,颔囊下长着很多下垂的毛发。

Last Refuge for Chinese Moose

Jian Ke works as a ranger for Genheyuan Wetland Park, so he is very familiar with the animals of the forests and wetlands. One of the animals he is most concerned about is moose. Give him a chance and he will tell you all about their habits. It is one of his favourite topics.

Moose are the largest members of the deer family and one of the largest land mammals in China. They live in Daxing'anling, Xiaoxing'anling and maybe still in Altai Mountains. They are still found in Daxing'anling in northeast of China and a few may still remain in the extreme northwest of China in the Altai Mountains. Adults may stand as tall as 2 m high. Their huge bodies are covered with short dark brown fur. They have large ears and a long square face. A flap of skin known as a bell sways beneath the moose's throat with a tassel of hair.

驼鹿身体硕大，胃口也很大。你难以想象出一只驼鹿一天要吃多少食物！驼鹿夏天每天可以轻易地吞食约30千克的食物，冬天每天能吃12千克。驼鹿吃树枝、树皮、根茎和木本植物的枝叶。它们喜吃柳树和白杨。在夏天，驼鹿还会吃水生植物，如睡莲、眼子菜和木贼类植物。在冬天，驼鹿就吃针叶树（常绿树）。驼鹿没有上犬齿，这方便它们在水底下啃食时把小型植物吸入嘴里。

驼鹿并不胆小，因为它们体形庞大，天敌很少。一群狼或一只棕熊打不过一只健康的成鹿，所以棕熊和狼通常会挑选幼小的、生病的或是年老的驼鹿作为猎捕对象。它们并不能直接杀死驼鹿，而是咬伤驼鹿造成伤口感染，这样驼鹿就会在几天内死亡。

一只死驼鹿会吸引众多的渡鸦叽叽喳喳争抢美食。它们的声音又会招来另一种吃腐肉的动物——貂熊。貂熊的嘴巴可是力大无比，一口就能在腐尸上咬下一大块。渡鸦是鸦科中最大的鸦，与普通的小嘴乌鸦相比，渡鸦的叫声更深沉也更刺耳，像大青蛙嘶哑的呱呱叫声。

雄驼鹿长有鹿角，它的作用主要是对战和炫耀。每年的9~10月，雄鹿都会大声喧叫以吸引雌鹿。在碰到天敌时，雄鹿并不会像你想象的那样用鹿角来战斗；它们的首选武器是尖利的蹄。它们用蹄能把狼或熊踢伤。

每年冬天，雄鹿的鹿角都会脱落，但在下一年的春天和夏天又会长出新鹿角。新生长的鹿角上附有一层松软得像"天鹅绒"般的毛皮，这层毛皮会给鹿角提供血液和养分。秋天刚到，鹿角上的"天鹅绒"开始脱落，雄鹿将鹿角在树枝上摩擦，将"天鹅绒"蹭掉。这样到了10月，它们就有新鲜、干净的鹿角来竞争和炫耀了。

鹿角也是驼鹿年龄的重要标识。每年冬天，桨形新角长出：新角会长出枝杈，再分出眉枝和主干，变成整副角架。5~8岁的雄鹿拥有最大的角架。随着年龄的增长，鹿角会变形，威风逐减。

With large size comes a huge appetite. You can hardly imagine how much food moose have to eat. They are browsers and will casually devour about 30 kg a day in the summer and 12 kg in the winter. They eat twigs, bark, roots and the shoots of woody plants. They prefer to eat willows and aspens. In the warm months, moose feed on water plants, water lilies, pondweed, and horsetails. In winter, moose browse on conifers (evergreen trees). Moose have no upper front teeth. This space allows them to suck small marine plants into their mouth while grazing underwater.

Moose are not so shy and because of their large size have few natural enemies. A pack of wolves or a brown bear is no match for a healthy adult moose, so bears and wolves typically can only pick off the young, sick, and old animals. A single bite can be enough to kill a moose as this is likely to cause an infection and the deer will die within a few days.

A dead moose attracts ravens that squabble noisily to get at the best bits. But the sound will attract another scavenger. The gulo gulo which has a powerful bite and can tear off and carry away large pieces of the carcass. Ravens are the largest of the crow family. They have a deeper and more rasping call than the common carrion crow and can sound like giant frogs croaking.

Antlers are only found on males, and used mainly for fighting and displaying. Bull males bellow loudly to attract mates each September and October. When fighting off predators, the antlers don't come into play as much as you would think; a moose's first line of defence is its sharp hooves, which are capable of mortally wounding a wolf or bear.

Bull moose shed their antlers every winter and grow new ones the following spring and through the summer. The growing antlers are covered in fluffy skin or 'velvet' which carries blood-flow supplying the antlers for as long as they are still growing. By early fall the bulls start to shed the velvet and shine their antlers by rubbing them against trees and by October they will have clean new antlers for competition and display.

Antlers are also a great indicator of age. With each winter, young moose paddles grow in size: nubs become spikes and spikes become full racks. Bulls in their prime, between ages 5 and 8, have the largest racks. With old age, the antlers become more deformed and less impressive.

雄鹿通常独处，但在交配的季节里，雄鹿会在很大的区域内寻找雌鹿。这时，三三两两的雄鹿会聚集在一起，用它们巨大的鹿角与其他小雄鹿群交战，迫使它们离开这片区域。获胜的雄鹿获得与雌鹿交配的权力，并筑建一片繁殖领地。交战的情景也不是你死我活的战斗，通常是其中一只驼鹿看到挑战者的鹿角更大、更炫时就会退出离开。当然，有大的鹿角也不是寻找伴侣的唯一途径——有些雄鹿方向感较好，或只是纯粹的运气好——可能偶遇到一只雌鹿，这样它就能完全避免鹿角战啦。

驼鹿全天都很活跃，特别是在黎明和黄昏的时候。驼鹿擅长游泳，它们常常待在溪流或湖水周围。它们能快速奔跑，速度可达每小时55千米。它们稳步走动的速度也能达每小时30千米，这比一般人都跑得快。

The bulls are normally solitary (live alone), but during mating season, bulls are very active and will cover a lot of ground looking for other females to mate with. A few bulls may come together, using their huge antlers, to battle with other males and fight them off from the area. The victorious male gets to mate with the female and to establish a breeding territory. The fights are rarely fight-to-the-death affairs, normally one competing moose will back away from a fight if the challenger has a more impressive rack of antlers or a harder kick. Great antlers are not the only way to find mates; some males with better navigational skills, or just sheer luck, may come across a female by chance and completely skip antler combat.

Moose are active throughout the day with activity peaks during dawn and dusk. They are good swimmers and can be found in and around streams and lakes. They move swiftly on land—running at speeds up to 55 km per hour. They are able to trot steadily at 30 km per hour (which is faster than the average man can run).

驼鹿独居独立,但有时两三只驼鹿会在同一条溪流觅食。小鹿常常紧随雌鹿。虽然罕见,但偶尔很多驼鹿也会聚集一起。汗马国家级自然保护区就曾经拍摄到一个画面,在冻结的湖上同时出现了11只驼鹿。

驼鹿会发出各种各样的声音。雄鹿会大声吼叫,也会发出嘎嘎叫声,或在发情期间发出吠叫声。雌鹿会发出长长的颤抖的呻吟声,以咳嗽般的声音收尾。雌鹿与它们的小鹿沟通时会发出咕噜声。雄鹿和雌鹿激动时都会发出哼哼响声。

雌驼鹿通常在每年的5月产下1～2头小驼鹿。小驼鹿出生时一般体重约15千克。刚出生的小驼鹿不会跑也不能自卫,但它们成长得非常快。雌鹿会和小驼鹿一起生活大约一年半,以保护她的孩子不被恶狼和棕熊吃掉。小鹿紧随雌鹿,母子情深。

Moose live alone and move around independently. Two or more individuals sometimes can be found feeding along the same stream. The mother and the calf form a strong social bond. In rare occasions more moose may gather. One video sequence filmed in Hanma National Nature Reserve shows 11 moose together in one herd on a frozen lake.

Moose make all sorts of funny noises. Bulls make a loud bellow, and they also croak and make a barking noise during the rut. Female moose, or cows make long quavering moans that end with a strange cough-like sound. They also grunt to communicate with their calves. Both sexes snort when agitated.

The cows, generally have 1 to 2 calves in May. On average, the calves weigh about 30 pounds at birth and grow very quickly. Still, baby moose don't have the ability to run or protect themselves very well, so the mother stays with her offspring for a year and a half, fighting off any wolves and bears that might try to pick off the young calves. The mother and the calf form a strong social bond.

驼鹿天生就是极好的游泳者，即使是新生驼鹿也会游泳。健科说他常常能看见驼鹿一跳入湖，就快速地往对面游去，估计能有每小时10千米的速度。即使新生驼鹿也会游泳。

以前大兴安岭到处都有驼鹿，数以千计。但是由于驼鹿的体形很大，视力又不那么好，它们很容易被人猎杀，因而数量剧减。健科说他有过撤除偷猎者的套子和夜里睡觉的藏身篷的经历。

20世纪80年代以来，内蒙古的乌玛、汗马、呼中等地区建立了国家级的自然保护区，这些地方成为驼鹿最后的庇护所。现在，已经禁止砍伐森林，新的自然保护区和湿地公园也纷纷建立，而且大兴安岭的大部分地区也都禁猎了，驼鹿的数量正在逐渐恢复。目前，集中在保护区内的驼鹿有好几百只，如果再加上分散在大兴安岭各处的剩余的鹿群，有可能达到一两千只。随着保护的深化，我们未来可能会看到越来越多驼鹿。但是，它们必须从保护区内分散出去。让人担心的是，越建越多的道路网络对迁徙的鹿来说是相当危险的。在健科看来，人们应该建立安静的地下通道和特殊的路口，让驼鹿可以安全地穿过公路和铁路。呼伦贝尔草原就在道路和铁路下专门为牛羊等动物建立了通道。

Moose are naturally gifted swimmers. Jian Ke tells us that it is common to see one hop right into a lake and swim across at up to 10 km per hour. The animals have an innate ability to know how to swim, so even calves can swim.

In former days there were moose all over Daxing'anling and their numbers must have been several thousands. But these animals are such good size and their eyesight is not so good. They are too easy to be hunted and so their numbers plummeted. Jian Ke describes how he used to work in dismantling wire snares and destroying the night shelters of poachers.

Wuma, Hanma and Huzhong were established as nature reserves since the 1980s and became the last refuges for these magnificent animals. But now the logging industry has been curtailed, new nature reserves and wetland parks have been created and hunting is controlled over most of Daxing'anling. The moose population is building back. There are again several hundred animals and maybe a couple of thousand if all the scattered remnant herds as included. With continuing conservation we may see these animals more often, but they need to spread out from their remnant strongholds and the growing network of roads will prove to be quite a hazard for migrating deer. Jian Ke thinks the road makers should build quiet underpasses and special crossing zones where they can safely get across the roads and railways, just like they do make for cattle and sheep to cross in the grasslands of Hulunbuir.

小测试

1. 中国在哪儿能找到野生驼鹿？

2. 雌驼鹿长角吗？

3. 驼鹿多长时间能长出新角？

4. 驼鹿吃什么？

5. 一只母驼鹿一年能生几只小驼鹿？

6. 驼鹿会游泳吗？

7. 什么动物能杀死驼鹿？

8. 小驼鹿跟随母驼鹿生活多长时间才能独立？

9. 雄驼鹿怎样找到交配的对象？

答案

1. 大兴安岭、小兴安岭，或许阿尔泰山也有；

2. 雌驼鹿没有角；

3. 雄驼鹿每年都会长新的角；

4. 柳叶和杨树叶、嫩树枝、水草，冬季吃松枝和各种草；

5. 每年最多两只小驼鹿；

6. 是的，它们非常擅长游泳；

7. 狼和棕熊会试着捕捉并吃小驼鹿；

8. 小驼鹿和雌驼鹿一起生活大约一年半；

9. 雄驼鹿发出深沉的叫声，并炫耀它大大的鹿角，吓跑其他雄驼鹿以吸引雌驼鹿与其交配。

Quizz Time

1. Where in China can you find wild moose?

2. Does the female moose have antlers?

3. How often do moose grow new antlers?

4. What do moose eat?

5. How many young can a female have in one year?

6. Can moose swim?

7. What predators can kill and eat moose?

8. How long does a calf moose stay with its mother?

9. How does the male moose find a mate?

Answers

1. Daxing'anling, Xiaoxing'anling and maybe still in Altai Mountains ;

2. The female moose does not have antlers;

3. Male moose grow new antlers each year;

4. Willow and poplar leaves and young twigs, sweet water grasses and in winter pine shoots and various herbs;

5. Up to two calves in any one year;

6. Yes, they are very good swimmers;

7. Wolf and brown bear would try to catch and eat young moose;

8. Calves stay with their mother for at least 1.5 years;

9. The male gives deep mating calls and show off its huge antlers that scare away rivals and attracts females.

乌林鸮怎样捕食

一天,巴拉和他父亲安岛拿了驯鹿角做的饰品到镇上去换了些盐和食物。傍晚时拎着这些东西回家时,巴拉很惊奇地看到一只灰色的硕大的鸟站在一棵小落叶松树的树顶上,黄色的眼睛含着些许意外,直直地看着他们。

"哇!爸爸,那是什么?"男孩问。

"孩子,那是乌林鸮。让我来给你讲一讲。"

How the Great Grey Owl Catches Its Prey

When Ba La and his father An Dao walked home from the town one evening carrying the salt, some food that they had exchanged for their reindeer horn carvings. They were surprised to find a large grey shape sitting in the top of a small larch tree, gazing directly at their approach with startling yellow eyes.

"Wow! What is that father?" asked the little boy.

"That my boy, is the Great Grey Owl. Let me introduce you a bit."

乌林鸮主要分布在北美洲、欧洲和俄罗斯的北部森林里。在中国，它们只分布在大兴安岭。它们生活在森林边缘和森林空地，以田鼠、老鼠为食，在开阔的平原上也猎食鼠兔。

"乌林鸮这么大！它一点儿都不怕我们！"巴拉大叫。

"是的，乌林鸮是世界上体形最长的猫头鹰，但并不是最重的。雕鸮和雪鸮都比乌林鸮重。乌林鸮看起来很大，主要是它的羽毛蓬松。如果除去了羽毛，乌林鸮的身体是相当瘦小的！"

然而，乌林鸮却是食量惊人的食肉动物。冬天，要在大兴安岭寒冷的环境中生存，乌林鸮每天要捕食6～7只老鼠或田鼠。夏天，如果在那乱糟的窝巢里有鸮娃娃的话，鸮爸爸还得捕捉更多的食物去喂食鸮妈妈和那些贪吃的鸮娃娃呢。

The Great Grey Owl is found scattered through the northern forests of North America, Europe and Russia. But in China it can only be found in Daxing'anling where it occupies forest edge and forest clearings and can feed on the voles, mice, rats and pikas in open country.

"The owl is so big! and he is not afraid of us at all!" exclaimed Ba La.

" Yes, the Great Grey Owl is the longest owl in the world but it is not the heaviest. The Eagle Owl and Snowy Owl are both heavier. The Great Grey Owl looks so big but that is mostly its puffed out feathers. Stripped off its feathers you would find a rather skinny little body inside!"

Nevertheless, the Great Grey Owl is a serious predator with a big appetite and need to catch about 6 or 7 mice or voles a day to survive in the cold climate of Daxing'anling. It may need to catch even more in the summer to feed Mrs owl and her greedy owlets if she has raised some young in her haphazard nest.

鸮爸爸要为全家狩猎,这意味着它要抓很多田鼠,这需要花费它很多的时间。所以鸮爸爸在清晨、傍晚和夜里的大部分时间里都在追捕猎物。

乌林鸮捕食时特别有趣。它暂歇在一棵小树、灌木或木桩上面,凝神贯注地观望着附近开阔的草地。突然间,它身体向前一倾,然后猛地俯冲下去。

有时候,它会暂时在猎物的上空徘徊一小会儿才猛冲下来从草底下把猎物抓住。

Mr Owl does most of the hunting for the whole family. That means a lot of hunting and takes a lot of time. So The owl hunts through the night and much of the morning and evening as well.

The Great Grey Owl has an interesting way of hunting. He perches on a low perch, a small tree, bush or man-made post and watches intently at the open grasslands nearby. Suddenly you will see him snap to focus, lean forwards then launch into a dive.

Sometimes he may hover briefly over some prey before diving down to grab into the grass below.

乌林鸮常常凭着感觉在草丛里抓上一爪子，其实它并没有看见里面有田鼠。有时候，它只抓上来一把草，不过更多的时候它的确能抓到田鼠之类的。它迅速地用嘴把猎物叼上。如果这是它自己的晚餐，它就会头一仰一口把猎物吞了。如果是更大的猎物，它会携带到一个方便的树枝，然后把它撕成碎片，再一块块地吞下去。但是，如果他猎到的是一家大小的食物时，它就会把猎物叼在嘴里飞到巢边，细心地喂给鸮妈妈和嗷嗷待哺的鸮娃娃们吃。

冬天，当地面上覆盖着厚厚的积雪的时候，乌林鸮还在忙着狩猎。这时它更了不得了，它冲入厚厚的雪层，出来时爪子上就攥着一只田鼠。它是怎么在看不见的情况下抓住田鼠的呢？答案是，它有绝妙的听力。

乌林鸮的整个脸像个雷达盘，能接收到在草地上或冰雪下爬行或进食的动物发出的轻微的声响，然后将信号放大传送到耳朵中。鸮能够通过脸盘感知是这边还是那边的声音较强，进一步确定声音达到一只耳朵跟到另一只耳朵之间的微小时间差，从而确定声源的位置。人类也能这么做，但是因为人的耳朵长在头的两侧，需要转动脑袋直到两边的声音达到平衡，才能确定声音的来源。

人类也能准确地判断出声音的来源，但是在确定声源的高度方面就不如鸮。那是因为人的耳朵长得一样高。但鸮的耳朵可不是！鸮的头稍有不对称，左耳比右耳略高。这使得鸮能从横向和垂直两个维度来确定声源。

乌林鸮既有脸盘作为声音放大器，耳朵又有三维方向感。虽然田鼠觉得躲在雪和草下面可安全了，但是乌林鸮还是能冲下去把它抓住！鸮甚至能发现30厘米深的积雪里躲藏的田鼠。

"那乌林鸮在哪儿筑巢呢？"巴拉问道。

安岛解释说："鸮妈妈个头比鸮爸爸还大一点。因为既要生蛋又要照顾小鸮，鸮妈妈也有很多工作要做。但她却很懒惰！她根本就不筑巢，只是把蛋生在乌鸦或鹫的老巢里，或是倒下的树干或树枝上的空洞里。它一连几年用同一个鸟巢。"

Mostly he grabs where he thinks there is a rodent without having seen it. Sometimes he comes up with nothing but a claw full of grass but often enough he has grabbed a vole and quickly transfers it to his big beak. If it is his own dinner he will toss it back in one mouthful. Larger prey he will carry to a convenient perch to tear off pieces and swallow bit by bit. But if he is hunting for the family he will fly off carrying the offering in his beak and land on the edge of the nest to offer delicately to his mate or directly to the eager chicks.

In winter when there is snow on the ground he is still busy hunting. Now even more remarkable he will plunge through the snow and come up with a vole in his talons. So how does he catch the mice he cannot even see? The answer is his remarkable hearing.

The entire face of the owl is like a radar dish, capturing and amplifying the slightest sound of a vole creeping or feeding under the grass or snow and transferring that signal to its ears. The facial dish provides accurate direction allowing the owl to pinpoint the sound exactly. We can do the same because our ears are on different sides of our head. We are able to judge if the sound is stronger or weaker on one side or the other and then measure the tiny time difference between the sound reaching one ear quicker than the other. So we can turn our head till the sound balances on either side and know that the sound is now coming from straight ahead.

We are quite good at lateral direction but not much good at then determining the height of a sound source. That is because our ears are at the same height. But not the owl! The owl's head is slightly asymmetrical. Its left ear opening is set slightly higher up its head than the right ear. This allows the owl to pinpoint sounds both laterally and vertically.

So the owl has both sound amplifier and great 3D direction finder and can pounce on the voles that feel so safe under their roof of snow or grass!! The owls can detect voles as deep as 30cm under the snow.

"So where do the owls nest? "asked Ba La.

An Dao explains further "With the job of producing eggs and looking after the small chicks, the female owl has a big job too and she is slightly bigger than her hunting husband. But She is a lazy nest builder! She does not make a nest at all but lays her eggs in the old nests of crows or buzzards or in the hollow of a broken tree trunk or branch. She may use the same nest for several years."

夏末，鸮娃娃羽翼未丰，但却跃跃欲飞。它们会爬出鸟巢，在附近的树枝上攀爬、跳动，嗷嗷叫着要吃的。这时鸮爸鸮妈必须昼夜奋战不停猎食，只有喂饱了鸮娃娃们，它们才得以安宁。这是一场与时间的比赛。因为寒冷的冬天即将到来，除非它们能够使鸮娃娃们身体长得足够壮硕，否则娃娃们可能会无法存活。因为在冬天，大兴安岭晚上的温度可能降到零下40℃以下。

乌林鸮通过竖起全身柔软的羽毛来保持温暖，但它们仍然需要大量的食物。鸮娃娃们还没有学会狩猎，所以鸮爸鸮妈整个冬天都会继续喂养娃娃们。同时，鸮娃娃们也慢慢学会自己狩猎，而且技术越来越熟练，直到它们能够听出那些美味的小田鼠隐藏在哪里。

"你看，乌林鸮并不怕人。你可以轻轻地走到离它只有几米远的地方，而它也就用有些惊讶但又无所谓的目光盯着你。"安岛说。

这时，乌林鸮很不情愿地拍打了几下翅膀，然后慢悠悠地滑翔到几十米以外的另一根树枝上停住了。乌林鸮飞行时悄无声息——这是它们的另一特长，这样鸮能够猛然无声地出击，等猎物注意到乌林鸮时已经太晚了。

"如果你有机会捡到一支乌林鸮的飞翼，你会发现它非常柔软蓬松。这样在飞行时羽毛之间的摩擦不会发出任何声音。乌林鸮虽然不太在意有人在附近走动，但对于也在大兴安岭越冬的，可能会与乌林鸮产生竞争的其他种类的鸮，比如长尾林鸮和鹰鸮，乌林鸮就会驱逐它们。"安岛说。

"乌林鸮的的确确是这林中的夜之王啊！"

巴拉和他父亲不知不觉中就走进了暮色中的林子。

"咱们走快点吧，现在天黑得早，还有很远的路呢。"

By late summer the baby owls are fluffy fledglings. They climb out of the nest space and hop about in nearby branches waiting for more food. Both parents now work on constant hunting to try to fill them up and keep them quiet. It is a bit of a race against time. Cold winter is approaching and unless they can get the new fledglings up to a large enough size, they will be unlikely to survive the temperatures that can drop to 40 °C below zero at night.

The owls can keep warm by fluffing out their soft feathers but they still need plenty of food and the youngsters are not yet very adept at hunting so both parents carry on feeding them through the winter whilst the new owls slowly learn to hunt on their own and get more and more skillful and hearing where those tasty little voles are hiding.

"As you can see, Great Grey Owls are not afraid of people. They may allow you to creep within a few metres, gazing at you with an astonished but rather indifferent gaze." continued An Dao.

Finally and reluctantly the owl gives a great flap and with a few leisurely slow wing beats glides to settle on another perch a few dozen metres further on. The flight is totally silent. This is another adaptation to help the quiet hunter attack its prey without their noticing until it is too late.

"If you are lucky enough to find the wing feather of the owl you will notice it is so soft and slightly fluffy. It is designed to slide over the next feather without friction or thus no sound. The Great Grey Owl may not care too much for people walking nearby but it will chase away the competition of other owls such as long tailed wood owls and smaller northern hawk owls that also spend their winters in Daxing'anling." An Dao Said.

"The Great Grey Owl is truly the king of the night birds here."

Bal La and his father walked on into the embrace of the dark woods.

"Better hurry up. It is getting dark so early these days. Still a long way to go."

小测试

1. 乌林鸮的公鸮还是雌鸮个子更大？

2. 乌林鸮在哪儿筑巢？

3. 乌林鸮每年都筑新巢吗？

4. 乌林鸮的主要食物是什么？

5. 乌林鸮的耳朵有什么特别之处？

6. 乌林鸮的眼睛是什么颜色？

7. 乌林鸮和长尾林鸮哪个更大？

8. 乌林鸮冬天会迁徙吗？

9. 乌林鸮能猎捕到多深的积雪下的食物？

答案

1. 雌鸟更大；

2. 在折断的树干上，或是鸦类的老巢里；

3. 不，它们只捡用二手鸟巢或用旧的巢；

4. 田鼠及其他小型哺乳动物；

5. 它们的双耳位置不对称，这样能识别声音的高度；

6. 亮黄色的；

7. 乌林鸮；

8. 不，它们全年都在大兴安岭觅食；

9. 积雪下30厘米处。

Quizz Time

1. Which is bigger the male Great Grey Owl or the female?

2. Where does the Great Grey owl nest?

3. Do the owls make a new nest each year?

4. What is the owl's main food?

5. What is special about the ears of the owl?

6. What colour are the owl's eyes?

7. Which is bigger Great Grey Owl or Long-tailed Owl?

8. Does the Great Grey Owl migrate for the winter?

9. How deep can the owl find food underneath the snow?

Answers

1. Female is bigger;

2. In the snapped tree trunk or an old crows nest;

3. No they use second hand nests or old nests;

4. Voles and other small mammals;

5. They are arranged asymmetrically to allow 3D hearing;

6. Bright yellow;

7. Great Grey Owl;

8. No, it hunts in Daxing'anling all year round;

9. As much as 30cm under the snow surface.

狐狸和貂熊

一天,狡诈的狐狸在林子里走动,它听到了一群渡鸦在叽叽喳喳争执不休。"那可是我的早餐。"狐狸想:"我得赶紧去看看发生了什么事。"

果然,有几只渡鸦聚集在一只刚刚死去的马鹿躯体旁。这只鹿一定是昨晚冻死的。渡鸦正争来争去看谁可以开吃第一口,可事实上它们谁也没有足够的力气啄开死鹿的皮,它们还得等待一些更大的动物把死鹿撕开后才能真正享受它们的盛宴。

狐狸恶狠地尖叫一声冲上去,渡鸦受到惊吓,猛然飞起,在死鹿的上空盘旋,愤怒地嘎嘎尖叫着,最后落到附近几棵树上,等待狐狸咬开死鹿的躯体。狐狸闻了闻看鹿肉是不是新鲜,随便咬了咬,但是鹿太大了,一点儿也咬不动。

现在狐狸遇到了麻烦。它要是咬开一只雪兔——它最喜欢的食物——一点儿也没问题。但是,这鹿太大了,怎么才能把它弄回去喂它那一大群饥肠辘辘的小狐狸呢?狐狸不得不绞尽脑汁,想出一个极佳的办法。

它可不希望棕熊这时出现。棕熊一来,就会把整只鹿都吃光,狐狸和乌鸦就一口都吃不上了。

The Fox and the Gulo Gulo

Once upon a time a clever fox was walking through the forest when it heard the sound of squabbling ravens ahead. "That could be my breakfast." thought the fox and he hurried up to see what was going on.

Sure enough several ravens were gathered around a freshly dead roe deer. It must have died of cold in the night. They ravens were fighting over who should have first pecks but in fact none of them had been strong enough to break through the skin of the dead deer and they would have to wait for some larger animal to tear into their feast before they could really get going.

The fox rushed up with an evil scream and sure enough the flustered ravens fled shrieking to circle around giving angry croaks and eventually settled in several nearby trees to wait for the fox to open up the carcass. Fox had a sniff to check the meat was fresh and gave the deer a preliminary tug but it was too big to move.

Now fox had a bit of a problem. He was big enough to carry off a whole snow hare—his favorite food. But the deer was far too big to carry home to his family of hungry cubs. Fox had to think what best to do.

He certainly did not want a brown bear to arrive. In that case the bear would take the whole deer and the fox and ravens would all get nothing.

狐狸想出一个办法。它赶紧过去找它的老对手貂熊。它知道貂熊的窝在哪里，就坐在貂熊居住的洞穴的门口，噪噪叫。

里面传来一咕噜声。貂熊从洞穴里走出来，满脸怒气。

"狐狸你到底想要干什么，不让我好好睡一觉？"

狐狸走到貂熊跟前，说："你嗅嗅我的嘴唇，貂熊，你闻到一头刚死的鹿的肉香了吗？"

确实，貂熊已经对鹿甜美的肉味垂涎了。

Fox had an idea. He hurried off to find his old enemy gulo gulo. He knew where the gulo gulo had a den and sat outside the tunnel entrance and gave a howl. There was a grunting from inside the den and finally gulo gulo appeared at the entrance looking rather angry.

"Whatever do you want fox. Can't I enjoy a well-earned sleep."

Fox came close and said "Can you smell my lips, gulo gulo? Do you smell the flesh of a freshly dead deer?"

Indeed gulo gulo was already salivating at the sweet smell of roe deer.

"狐狸，鹿在哪里？"貂熊问道。

"跟我来，我们来一个小小的比赛，看看这片林子里，到底谁才是真正最聪明的吃货。"

貂熊怀疑狐狸又在耍小伎俩，但鹿肉太美味了，就不再跟它争辩。貂熊抬起脚一拖一拽地，随着狐狸快快过去。

这事狐狸太清楚了，貂熊个子更大，但它更有一张强劲有力的嘴巴。它才有能力把鹿咬开口。狐狸盘算着，要怎样说服貂熊将鹿撕咬成块，却又万万不能让貂熊把整条鹿给弄走吃掉。

狐狸和貂熊赶到死鹿身旁，把两只回来再试试运气的渡鸦赶走，渡鸦早已设法将死鹿的眼睛啄了出来，现在更美味的肉味已经在晨曦中弥漫。

"貂熊，我们这样比赛吧。"狐狸跟头脑简单的貂熊说："鹿是我找到的，所以我可以先开口吃，谁能撕下最大一块肉就赢了，谁就可以把整条鹿带回家，好吗？"

"好嘞。"貂熊回答说。"我想我这一次一定会赢的。"

狐狸第一个上去，一口咬住死鹿又长又软的耳朵，费力地拉啊曳啊。最后，好不容易撕下鹿的一片耳朵。它将耳朵放在鹿旁边，往后一站。

"好，轮到你了。"狐狸对貂熊说。

貂熊觉得狐狸的功夫也太可悲可笑了。它拖身上前，一口狠狠地把牙深深扎进了鹿的后腿，把这可怜的死鹿又拖又撕，越来越用力，因为死鹿露出来的肉强烈地刺激着它的嗅觉。

随着最后的一声咕噜，貂熊一曳把一整个鹿腿撕开，貂熊把这条鹿腿放在被狐狸咬下来的一小片鹿耳朵旁边。

"Where is it, fox?" said the gulo gulo.

"Come with me and we can have a little competition. We can see who is really the cleverest carnivore in this end of the forest."

The gulo gulo suspected this was another of fox's little tricks but the smell of the deer was too sweet to argue. He hurried along after the fox with his shuffling quickest walk.

Now what fox knew only too well was that the larger gulo gulo had a famously strong bite. He was certainly big enough to rip open the deer. The trick would be to persuade the gulo gulo to break the deer into bits and not to run off with the entire corpse.

When fox and gulo gulo got back to the dead roe deer they had to chase off two of the ravens that had already returned to try their luck again. Indeed the ravens had already managed to peck out the deer's eyes and an even sweeter smell of the meat within spread on the morning air.

"Now gulo gulo, here is the competition." explained the fox to the simple gulo gulo. "I found the deer so I have first go". "Whoever can tear off the biggest bit of meat is the winner and can take the deer home. OK?"

"OK." replied the gulo gulo. "I think I will certainly win this one."

Fox took his first turn and went up to the dead deer. He took a firm bite of the long tender ear of the deer and tugged this way and that for all he was worth. Eventually a piece of the ear tore free and fox placed the piece of ear next to the deer and stood back.

"OK. your turn." he said to the gulo gulo.

Gulo gulo thought the fox's effort was pathetic. He shuffled across sank his teeth deep into the deer's back thigh and started tugging and tearing at the poor animal getting more and more aggressive as the taste and smell of the exposing flesh enraged his nostrils.

With a final grunt and tug the entire leg tore in half and the gulo gulo laid down the ripped leg next to the fox's severed ear.

"认输吧,狐狸,我公平合理地赢了你。"貂熊说。

"别急,朋友。我们俩都把鹿撕成了两块,都赢了。你动作粗暴,把鹿弄得乱七八糟,只剩下三条腿,而我却更细致,留下了一整条美味的鹿,我当然应该得大部分。"

"你这个骗子狐狸!"貂熊愤怒地咆哮着。

这时,渡鸦都看到它们吃食死鹿的机会。它们尖叫着狂冲下来,疯狂乱啄一阵,快快吞食了最美味的小块肉。

"Give up fox. I won that fair and square." said gulo gulo.

"Not so fast my friend. We both succeeded in tearing the deer into two pieces. You made a crude mess and left only a mangled three legged corpse. I was much more delicate and left an entire tasty deer. My section is certainly the larger."

"Why you cheating little dog." roared the angry gulo gulo.

But by this time the ravens could see their way into the corpse. They descended in a screaming frenzy pecking hither and thither removing and gobbling the most tasty morsels.

"这么着吧,老朋友,"狐狸说:"虽然你没有充分理解我们的规定,我也就让了你,把大份的给你,而我就只分那只你撕咬下来又扔掉的烂腿。"

就这样,狐狸叼起那条大小正好的马鹿腿,赶紧回到它的洞里去喂小狐狸。而貂熊呢,它还在马鹿边大怒不止。这时候,渡鸦见到马鹿被撕开口子,也壮起了胆,蜂拥着回来啄食马鹿肉。貂熊又得与一大群渡鸦开战争肉吃了。

"Tell you what old mate." said fox, "As you did not understand the rules properly, I'll let you keep the biggest piece of the deer and I will content with this mutilated leg you have discarded."

So fox picked up the perfect sized piece of tasty leg and hurried off to his den and waiting cubs, leaving the still angry gulo gulo to fight over the rest of the roe deer with the emboldened flock of great black crows.

小测试

1. 为什么狐狸不能把整条死鹿都留给自己享用?

2. 渡鸦为什么不能自个吃马鹿?

3. 为什么狐狸要去找来一只危险的动物来帮忙?

4. 貂熊住哪儿?

5. 狐狸和貂熊,谁的嘴巴更强劲有力?

6. 如果棕熊来了,会发生什么?

7. 狐狸和貂熊,谁更聪明?

8. 为什么狐狸只拿走一块肉?

答案

1. 马鹿太大,狐狸自个儿咬不开口也拖不动;

2. 渡鸦啄不开马鹿的厚皮;

3. 它必须去找一只有足够的力气把马鹿厚实的皮咬开口子的动物,它必须冒这个险;

4. 住在洞穴里;

5. 貂熊更强劲有力;

6. 棕熊来了会把整只马鹿都弄走;

7. 这次狐狸显得更聪明;

8. 因为这块肉大小正好,它能把肉叼回到洞里。

Quizz Time

1. Why cannot the fox keep the whole deer for himself?

2. Why cannot the ravens eat the roe deer on their own?

3. Why does he choose a dangerous animal to help him?

4. Where does gulo gulo live?

5. Which has the stronger bite, fox or gulo gulo?

6. What would happen if a bear appeared?

7. Which is the cleverest animal, fox or gulo gulo?

8. Why does the fox take off the smaller piece of meat?

Answers

1. The deer is too large for a fox to open up or carry away;

2. They are not strong enough to break through the tough skin of the deer;

3. He has to get help from an animal strong enough to break up the dead deer, he has to take a risk;

4. In a hole den;

5. The gulo gulo is much stronger;

6. The bear would steal the whole dead deer;

7. In this case the fox shows he is cleverer;

8. It is just the right size to carry back for his cubs.

松芽早餐——黑嘴松鸡的故事

"妈妈,午饭吃什么?"松鸡小喳叫道。"好吃的落叶松嫩芽。"松鸡妈妈回答道。"哦,妈妈,我们早饭、昨天和前天吃的都是落叶松芽,太恶心了,我们能不能换点好吃的东西?比如以前你抓的那些甲虫,怎么样?"小喳满脸沮丧。它的羽毛沉沉地、紧紧地贴在身上。

"小喳,之前我就告诉过你,昆虫是用来喂娃娃的,你和你的姐姐们早就长大了,现在已经是夏天了,我们必须吃云杉的芽和叶,你姐姐她们并不抱怨。你去吃点落叶松芽吧,否则,我就告诉你爸爸。"

一提到爸爸,小喳就心惊胆战。它的爸爸可是又大又吓人。幸亏,它爸爸老不在家。松鸡爸和它的伙伴们每天在一起跳舞、炫耀。

Pine Shoots for Breakfast—a Capercaillie Story

"What's for lunch Mama?" called Charlie capercaillie. "A nice dish of larch shoots." replied mother Capercaillie. "Oh Mama, we had larch shoots for breakfast and yesterday and the day before. They are horrible! Can't we have something decent to eat for a change? How about some of those nice beetles you used to catch?". Charlie looked dejected. His feathers sank to a tight limp bunch.

"I have told you before Charlie. Insects is only for baby chick capercaillies. You and your sisters are well beyond that stage. It is summer now and we have to eat spruce buds and shoots. Your sisters don't complain. Now eat up your larch or I will have to tell your father."

Mention of Charlie's father brought a tremor to little Charlie. His father was huge and scary. Luckily his father was never at home. He was off with his men friends dancing and showing off every day.

小喳的爸爸是这一带的舞蹈冠军,需要定期练舞。因为总是有一些年轻的雄松鸡试图跟它一比高低,想把冠军从小喳爸爸手上夺走。

黑嘴松鸡的舞蹈很有戏剧性。首先,公鸡尾巴直立,将嘴巴往上一伸,双脚在大树枝上跺来跺去,发出难以置信的颤音;随后,它们跳到林地里,挥舞着翅膀,继续踢踢踏踏地跳舞,同时发出奇异的叫声:呼,呼,咚咚咚咚咚……"

Father capercaillie was the champion dancer hereabouts and he needed to practice regularly. There were always a few younger males trying to compete and oust him as champion.

The capercaillie dance is a dramatic affair. First the cock birds erect their tails and extend their beaks high in the air and stomp about on large tree boughs making an unbelievable throbbing noise. Then they jump down to the forest floor with a fanfare of flapping wings and continue their strutting stomping dance and weird calls on the ground. Whirr whirr, tonk tonk tok tok tok.

因为小喳的爸爸是这里最高大的松鸡，比其他雄鸟高出一头，所以令它们望而生畏。但是常常也有一些雌松鸡会对它抛媚眼。小喳还听说过，可能还有其他小松鸡也管自己爸爸叫爸爸呢！

小喳索然无味地啄了啄它的午饭，看着两个姐姐顺从地咀嚼着绿松芽。小喳虽然还饿着，但它决定自己去走走，反正等会儿一家人还要在灌丛下偎依休息。小喳妈妈说过，一旦它们这些小松鸡们能飞了，自己就要花更多的时间在树冠上生活。

小喳朝着它熟悉的路走下坡，到了路尽头的空地时，它仍然感到踌躇满志，大步走进了以前从未见过的一片林子。

它尝了尝一个看起来比落叶松芽好吃的蘑菇，但蘑菇黏糊糊的，使它有点头晕。然后它看到一个大土堆，上面歪歪扭扭地爬满了棕褐色蚂蚁。小喳想起了自己还是娃娃时，妈妈曾经给它吃过好吃的小虫。它啄了一只蚂蚁，尝了一下，有点酸，蚂蚁还咬了它的嘴。更多的蚂蚁从土堆里跑了出来，开始咬它的脚和腿。小喳赶紧跑开，摇了摇腿，把蚂蚁抖掉后又继续往前勇敢地探索。

突然，小喳听到深沉的咆哮声，它吃惊地看到一群野猪在林地里拱来拱去。野猪们头朝下，似乎正在吃东西。它们看起来很大，也非常凶猛，它们有大獠牙、灰白毛发，小喳躲着不敢被它们看见。

野猪过去后，小喳才敢出来看看它们在那里做什么。原来它们在地上拱来拱去，把土块、树根拱得到处都是。它们是不是在吃泥巴？要不它们在做什么？小喳觉得野猪真是脏啊。

As father was the biggest bird he could tower above his rivals and scare away all the other cocks but he usually had a few interested glimpses from the hen birds who hung around admiringly. Charlie had heard stories that his family may not be the only ones to call Father-Baba!

Charlie pecked with little interest at his lunch, and watched his two boring sisters dutifully gobbling up their own greens. Then still feeling hungry Charlie decided to wander off on his own whilst the rest of the family snuggled down in the underbrush to rest. Mama had told the children that as soon they could fly then they would live more time up in the tree tops.

Charlie headed down the trail he knew quite well but when he came to the clearing at the end of the trail he was still feeling very bold and headed on further into a new sector of the woods he had never seen before.

He tried tasting a mushroom that looked rather more interesting than larch leaves but it was slimy and made him a little dizzy. Then he found a big mound writhing with active brown ants. He remembered the nice beetles his mother used to feed him when he was a chick. He pecked up one ant and ate it but it tasted acid and bit him in his mouth. Many more ants came running from the mound and started biting his feet and legs. Charlie ran away, shook his legs to get the ants off, then carried on his brave exploration.

Charlie was alarmed to hear deep grunting ahead and was surprised to see a group of wild pigs working through the forest. They were busy nose down and seemed to be eating something but they looked very big and very fierce with big tusks and hoary beards and he did not dare show himself.

When the pigs had passed he came out to see what they had been up to. What a mess they had made digging about in the soil tossing small plants hither and thither. He could not see what they had been so interested in unless they eat mud. What dirty animas he thought.

小喳继续往前走。前面的树木更稀疏，可以看到树木之外有更广阔的一片空地。小喳很惊讶地看着。这就像一个崭新的世界。它可以听到很多鸟儿快乐地歌唱。那里肯定有好吃的东西。

小喳看到一群鸟儿在草地上啄食。它们长得跟小喳见过的其他黑嘴松鸡很像，就是个子小一些，也没那么吓人。

它们肯定在吃什么东西，小喳大胆地朝它们走去。"你们在吃什么？"小喳问离它最近的松鸡——事实上，这是黑琴鸡一家。

"我们在吃石南花的叶子，很好吃。"受到鼓励，小喳也啄吃了一点儿这又老又干的叶子。它试着吞下，但很快就吐了出来。"哦，好恶心啊。"它尖叫道。

"是的，石南叶味道很重，但我们吃习惯了。"琴鸡朋友说。"你可以尝尝那边的越橘叶。"

小喳这次对越橘叶更谨慎些。越橘叶味道还好，也许越橘叶可以换一换可怕的落叶松芽。它又多吃了几片叶子。

"不要吃太多的越橘叶，我们需要它们在秋天时产果子，越橘果是我们最喜欢的食物。"琴鸡说。

"好的，我会记住的，我最好现在回家去了，谢谢你们的陪伴，希望很快再见到你们。"

小喳原路返回，看到自己一家人刚从午睡中醒来。

"妈妈，我们这段时间可以吃越橘叶吗？"小喳天真地问。

"亲爱的，过几个星期，就会有很多浆果和水果可以吃，越橘、蓝莓、覆盆子、野草莓。那时你真的可以吃到好东西，你就会长得像你爸爸一样高大。事实是，你也需要吃很多东西，因为在冬天，这里会变得很冷，有时甚至没有任何落叶松芽可以吃呢。"

Charlie wandered on. The forest was thinning and he could see a wide open space beyond. Charlie gazed out in amazement. It was like a bright new world. he could hear lots of birds singing happily. Surely there must be good food out there.

As he watched he saw a family of birds pecking around in the grass tussocks. They looked really familiar—really like the other capercaillies he had met, but there were quite a lot smaller and less frightening.

They were definitely feeding so Charlie was bold enough to walk towards them. "What are you eating?" asked Charlie as he reached the nearest of the grouse. For indeed he had discovered a family of black grouse.

"We are eating heather leaves. They are good." Encouraged, Charlie pecked a few of the leathery leaves and tried to swallow them. He spat them out quickly. "Oh they are disgusting." he squawked.

"Yes they are quite strong, but we get used to them." stated his new grouse friend. "You could try the bilberry leaves over there."

Charlie was more cautious with the bilberries. They were alright. Maybe they could make a change from the dreaded larch. He munched a few more leaves.

"Don't eat too much bilberries. We need them to produce fruits in the autumn. That is our favourite food." said the grouse.

"OK. I will remember that. I better go back to my own family now. Many thanks for showing me around and hope to see you again soon."

Charlie hurried back the way he had come and found the rest of his family just waking up from their midday snooze.

"Mama, can we have bilberries to eat some time?" asked Charlie innocently.

"Well my dear, in a few weeks there will be lots of berries and fruits to eat, bilberries, blueberries, raspberries, wild strawberries. then you can really have good food and grow as big as your father. In fact you need to eat a lot because in winter it gets very cold here and there are not even any larch shoots to eat."

小喳一想到吃水果的季节就感到兴奋,不过对冬天又有点担心。
"如果没有云杉叶,妈妈,冬天我们吃什么?"它担心地问。
"哦,别担心,亲爱的,冬天还早着呢,那时总会有大量的松叶。"
"啊!"小喳咕哝着。

Charlie was excited about the idea of a fruit eating season but was a little worried about winter.

"What do capercaillies eat in winter if there are no spruce leaves, Mama?" he asked concerned.

"Oh do not worry yourself yet my dear. Winter is still far away and there are always plenty of pine leaves."

"Ugghh." muttered Charlie.

小测试

1. 松鸡小喳早饭吃什么？

2. 为什么小喳没有它最喜欢的甲虫做早餐？

3. 小喳为什么怕它的爸爸？

4. 小喳的爸爸为什么花这么多的时间练舞？

5. 小喳的爸爸是个好的舞蹈家吗？

6. 黑嘴松鸡雄鸟和雌鸟有什么不同（两点）？

7. 为什么野猪在地上翻出这么多土？

8. 为什么小喳觉得自己跟黑琴鸡有点儿一见如故？

9. 黑琴鸡吃什么？

10. 黑嘴松鸡冬天吃什么？

答案

1. 落叶松的嫩芽；

2. 虫子是松鸡娃娃的食物，小喳已经长大了；

3. 它的爸爸又大又可怕；

4. 黑嘴松鸡雄鸟通过比舞竞争地盘和得到雌鸟的青睐；

5. 是，它的爸爸是舞蹈冠军；

6. 雄鸟体型较大，为黑色；

7. 它们可以通过嗅觉在土里找到虫子、坚果和蘑菇来吃；

8. 黑琴鸡与黑嘴松鸡是关系相近的两种松鸡物种；

9. 石南花的叶子以及一些水果和昆虫；

10. 落叶松和松树的枝条和嫩芽。

Quizz Time

1. What does Charlie get for breakfast?

2. Why does Charlie not get his favorite beetles for breakfast?

3. Why is he afraid of his father?

4. Why does his father spend so much time dancing?

5. Is Charlie's father a good dancer?

6. Describe two differences between male and female capercaillie's.

7. Why do pigs dig so much in the soil?

8. Why does Charlie feel a bit at home with the black grouse?

9. What do Black grouse eat?

10. What can capercaillie eat in winter?

Answers

1. Larch shoots;

2. Insects are only for the small baby chicks, he is too big already;

3. His father is so big and scary;

4. Cock capercaillies compete for territories and for females by dancing contests;

5. Yes, his father is the dancing champion;

6. Cock is much bigger and coloured black;

7. They can smell out worms, nuts and mushrooms to eat;

8. Black grouse are a related grouse species;

9. Leaves of blueberries plus some fruits and insects;

10. Twigs and buds of larch and pine.

傲窦的小狼

一天，小傲窦离开帐蓬要去把自家的小驯鹿带回蓝地。这两只小驯鹿离开营火和帐篷太远了，要是进入了茂密、幽深的森林会不会遇到什么危险呢？就在上个星期，有个养驯鹿的人就说，他的一只驯鹿被狼吃了。小傲窦决定去把小驯鹿给找回来。

傲窦能猜出来她的两只小驯鹿到哪里去了。肯定是去找甜越橘吃了。她随着驯鹿的足迹走，一边慢慢地走一边享受着初夏森林的美景。一棵高大的落叶松的树皮上有沙沙声，她抬头一看，有两只黑松鼠在树上蹿来跳去、你追我赶地玩着，然后又回到了筑在高枝上的球形窝巢里。

很快，傲窦就赶上了四处游荡的小驯鹿。她轻声地呼叫它们。它们抬起头，看到傲窦很高兴，都走过来让她拍拍头。

傲窦转身领着小驯鹿回去，这时她听到喵呜喵呜的低叫声。她走过去看看，很惊讶地发现一只小狼躲在一棵老树干的裂缝处。小狼看起来很痛苦，好像是饿坏了。傲窦环顾周围的林子，看看狼妈妈是不是在附近，但她没有找到。这小狼是被遗弃了吗？傲窦走过去仔细看了看小狼。她抱起小狼，一点儿都不害怕。她注意到小狼的左后腿受了伤，她把小狼放到地上，小狼用其他三条腿几乎站不起来。

傲窦看着小狼无助的小脸，决定把它带回家去。她把小狼裹进她上衣里，小狼并没有挣扎。傲窦叫上两只小驯鹿，把它们都一起带回了营地。

Aodu's Little Wolf

One day little Aodu left her tent and went to round up the young reindeer. They had wandered too far from the camp fire and who knows what dangers they could run into if they went deep into the tall forest? Only last week one of the herders reported he had lost a deer to wolves.

Aodu could see where the two young deer had gone in search of sweet bilberries. She followed slowly enjoying the sweetness of the early summer morning. She heard a rustling on the bark of a tall larch tree and looked up to see two black squirrels chasing each other energetically round and round the tree then back up their ball-shaped nest in the higher branches.

Soon she caught up with the adventurous young deer. She called them gently. They raised their heads, seemed happy to see her and both trotted forward to get a pat on the head.

Aodu turned to lead them home when she heard a low mewing noise. She went to investigate and was surprised to find a small wolf puppy hiding in a cleft in an old tree trunk. The puppy looked miserable and starving. Aodu looked around the forest in case the mother wolf was nearby but there was no sign of any other wolves. The baby had been abandoned. Aodu went to look closer at the pup. She was not afraid as she picked up the sick animal. She noticed it had a wound on its back left leg and when she put it down on the ground, the little pup could hardly stand on its other three legs.

Aodu looked at the sweet face of the helpless wolf pup and decided to take it home. The pup did not struggle as she gathered it up into her blouse, called to the two young reindeer and led her collection of animals back to camp.

看到她和驯鹿都安全回来，马妮德老奶奶很高兴。奶奶把小驯鹿带过去与其他驯鹿放在一起，在一堆营火旁边。夏天天转暖时，苍蝇和蚊虫很多，在营火旁边，驯鹿就不会被咬。

　　老奶奶看到傲窭衣服兜着的东西时，有些恐慌："哦，亲爱的，你不能留下这东西，你爸爸要是看见了会把它弄死的，你知道他有多讨厌狼。"

　　"可是，奶奶，你看它多可爱，又受了伤，我想把它当宠物，让它恢复健康。"

Grandma Ma Ni De was happy to see her return safely with deers. She herded them over to join the rest of the small herds gathered around the smoldering fire to escape from the biting flies and midges that were become a pest as summer warmed up.

The old grandmother looked in some horror at the pathetic bundle in Aode' blouse. "Oh my dear. You cannot keep that. Your father will kill it if he sees it. You know how he hates wolves."

"But grandma, it is so sweet and injured. I want to look after it as a pet and nurse it back to health."

毛茸茸的大狗已经闻出了营地来了不速之客，它一摇一嗅地朝奶奶和傲窦走来。通常，如果有新奇东西到来，大狗都会吠叫。但不知为什么，大狗并不觉得小狼很奇异，它只是想过去好好嗅一嗅。它已经接受了狼是家庭一员了。

傲窦兴高采烈地跟奶奶说了路上的事和她怎样找到了小狼。

奶奶觉得收养一只狼并不是什么好事，但她特别宠爱傲窦，很同情傲窦没有其他孩子一起玩。

"那好吧，你可以把小狼放在藏物棚里，里面很温暖，你每天都可以喂养它，你爸爸也不会看到，但我们必须教它保持安静。"

所幸的是，傲窦的爸爸大部分时间都不在家。他参加了一个伐木组，留下他母亲和可靠的大狗去照顾家人和20多只驯鹿。只有在周末，他才回到营地。大部分的时间他都在睡觉。这样，奶奶和傲窦设法秘密地把这只狼崽子藏在营地三个星期。

每天傲窦都会去看小狼，跟小狼玩儿。她会选一点食物或肉干给小狼吃，还用碗带些驯鹿奶给小狼喝。每天她给小狼的腿伤抹杜香油，绑绷带，小狼的伤很快就更好了。狼崽子现在看起来很快乐，它毛发蓬松，越长越大。傲窦也很开心，她织了一条漂亮的花围巾，戴在小狼的脖子上。

Dagou the big shaggy camp dog had smelled the newcomer and came snuffling up to the two females. He would normally bark at strange things but somehow the puppy did not seem strange and Dagou wanted a good smell. He had already accepted it as part of the household.

The excited girl told her grandmother about her walk and how she had found the wolf pup.

Grandma felt that no good would come of adopting a wolf but she loved Aodu very much and felt great sympathy for the little girl who had no other children to play with.

"All right. You can keep the pup in the small fur store. It is warm in there and you can visit and feed it each day. Your father will not see it but we must teach it to keep quiet."

Luckily Aodu's father was away most of the time. He had started a job helping the timber felling team and left his mother, daughter and trusted Dagou to look after the 20 reindeer that they family still owned. It was only at weekends that he came back to the camp. If Grandma let him have some wine he would sleep most of the time and the two womenfolk managed to keep the wolf cub a secret for three weeks.

Every day Aodu would visit the pup and play with it. She fed it choice food and pieces of dried meat. She brought it reindeer milk in a bowl. She bandaged the injured leg and rubbed fat and the sweet oil of Ledum leaves into the wound, each day. It soon got better. The cub looked happier now, fluffy and getting bigger. Aodu was very happy. She made a beautiful embroidered scarf for the cub to wear round its neck.

现在，小狼还能跟傲窦和大狗到林子里走走。即使小狼的牙齿非常锋利，傲窦也常常和小狼一起玩，小狼的玩法就是咬一咬她，扯一扯她。大狗似乎也喜欢小狼的玩法。

Now the cub would join Aodu and Dagou to go on walks in the forest. Aodu would play with the little wolf even though its teeth were really very sharp and its idea of playing was biting and tugging at her. Even old Dagou seemed to enjoy the cub's games.

但是，有天晚上，傲窦的爸爸回了家。深夜，他听到一声奇怪的嚎叫，于是寻声去看个究竟。果然，他发现小狼。爸爸一把揪住小狼的颈背，回到帐篷。

"这是怎么回事？"他叫道："我的斧子在哪？"

"别，别杀小狼，这是傲窦的宠物，她很爱它。"奶奶说。

But one night when father was back at home, there was a strange howling. Father got up to investigate and headed straight for the fur store from whence the calls seemed to come. Sure enough he found the little cub, grabbed him up by the scruff of the neck and carried him back to the living tent.

"What is this?" he demanded. "Where is my axe?"

"No. Do not kill the little wolf. It is Aodu's pet and she is much in love with it." called Grandma.

突然营火那边传来一声尖叫,爸爸跟大狗冲出去看是怎么回事。很快他们又回来了。

"两只狼想要吃小驯鹿。"他说:"还好我们及时赶走了它们。"

"你看,"傲窦兴奋地叫道:"是我那只小狼先感觉到了野狼来了就报警了。要是我们没醒来,我们就可能又失去一只鹿。幸亏我的小狼相救!"

爸爸这才心软下来,说:"好吧,准许你把狼当宠物养,不过它得老老实实,别招惹驯鹿。"

傲窦给了爸爸一个大大的拥抱感谢他。爸爸高高兴兴地回去睡了。小狼那天晚上就睡在傲窦床上。

就这样,小狼跟这家鄂温克族人一住就是两年,帮着大狗一起保护他们的驯鹿群。

第三年,野狼的嚎叫又在山上响起。小狼无法拒绝一只小母狼的气味,离开鄂温克族人一家,回归到原野,成立了自己的小家。

几年过去了,小傲窦还是没能忘了那只小狼。她跟奶奶说,她走在森林里找浆果和蘑菇时,还时不时地能看到狼。但是,没人知道这是不是她的想象,因为狼现在真的很少见了。

But then there was another scream from the deer by the campfire. Father rushed out again with Dagou in hot pursuit. Soon they returned.

"Two wolves were trying to get at the young deer." He announced. "We were just in time to chase them off."

"There you are." cried Aodu. "It was my wolf cub that sensed the wild wolves and raised the alarm. If we were not all awake we would have lost another deer. My puppy saved the day."

So father relented. "OK. You can keep your pet so long as he behaves and leaves the deer alone."

Aodu gave her father a thank you hug. The big man went back to his bed content enough. The wolf cub slept with Aodu for the rest of the night.

And so it was that the wolf stayed with the Evenki family for two years, helping Dagou to protect the small deer herd.

It was in the wolfs third year that the howls of wild wolves were heard again in the hills and the smell of a young lady wolf proved too irresistible. The young wolf left the Evenki family and went back to the wild to form his own small pack.

A few years later, he still can't forget the Wolf. She tells her grandmother that she still sees the wolf from time to time when she goes on walks in the forest to search for berries and mushrooms. But no-one knows if she is imagining it as wolves are really rather rare now.

小测试

1. 傲窦第一次看到受伤的小狼时,她是为什么到营地外面去的?

2. 为什么傲窦的父亲憎恨狼?

3. 狼跟人有什么相似之处?

4. 傲窦怎么喂养她的宠物小狼?

5. 狼妈妈为什么丢弃小狼?

6. 傲窦怎么救护了这只受伤的小狼?

7. 为什么这只宠物狼最终要离开傲窦的家?

8. 你相信傲窦还能认得出她小时候的宠物吗?

9. 你要是带回来一匹小狼,你父亲会说什么?

答案

1. 她去寻找两只迷失的小驯鹿,因为它们离开营地比较远;

2. 因为狼吃了他的小鹿;

3. 狼也是群聚动物,通常一大群在一起;

4. 她对待狼就像对家里的狗一样,给它喂食;

5. 可能是因为小狼受了伤,赶不上它的一家;

6. 她将小狼的伤口包扎起来,还用杜香油擦拭;

7. 因为小狼长大了,要找一只合适的小母狼做配偶;

8. 狼和人都有很长、很好的记忆力,他们可能可以相互认识;

9. 他可能会要我弄走它!

Quizz Time

1. Why did Aodu go out of camp when she first found the injured wolf cub?

2. Why does Aodu's father hate wolves?

3. Why would a wolf bond with a human houseshold?

4. How did Aodu feed her wolf pet?

5. Why did the wolf's mother have to abandon her cub?

6. How did Aodu treat the injured cub?

7. Why did the pet wolf eventually leave Aodu's family?

8. Do you believe Aodu can still recognize her old pet?

9. What would your father say if you brought home a wolf cub?

Answers

1. To look for two missing reindeer that wandered too far from camp;

2. Because they eat some of his young deer;

3. Wolves are social animals that need to be members of a pack;

4. She treated it like the family dog, fed it meal scraps;

5. Because the cub was injured and could not keep up with the family;

6. Bandaged the wound and rubbed it with aromatic oil;

7. He was old enough to want to mate and needed to find a suitable wild female;

8. Wolf and humans both have long social memory, and can probably recognize each other;

9. He would probably tell me to get rid of it!

天鹅之家

"那是什么声音?"小巴拉问。他听到一阵似鲲鹏展翅的呼呼声,还有成群的天鹅的咕噜声。"这就是春天的声音。"他父亲安岛说:"这是水禽往北回到它们传统的繁殖地时大举迁徙的声音。"

天鹅和豆雁在高高的天上翱翔,然后翻腾、滑落。它们来此小憩几天。"打开家门,作为东道主。我们应该好好善待我们尊敬的客人。"小巴拉的父亲说。

就这样,天鹅和豆雁从空中降落到了大兴安岭的农田和湖泊。

有些水禽会在湿地上、草地中筑巢,有些水禽则在这里歇歇,吃吃,然后继续往北,越过俄罗斯边界,飞到西伯利亚北极苔原上。

安岛给他儿子介绍了豆雁、大天鹅和小天鹅这三个来大兴安岭小憩的鸟种。

A Hotel for the Great Swans

"What is that noise?" asked Ba La. He heard the wumf wumf wumf of great wings beating, the honking of many geese, the muttering of flocks of great swans. "It is the sound of spring." said his father An Dao. "It is the sound of the great migrations of waterfowl heading north on their way back to their traditional breeding grounds."

High—so high—overhead they could see the flocks of swans and geese flying, and then yes they are turning, they are gliding down. They are coming to stay a few days. Open the hotel gates. "We must be good hosts to our respected guests." said Xiao Ba La's father.

And so the flocks spiral down from the heavens and land with thud or splash in the farm fields and lakes of Daxing'anling.

Some of the waterfowl will nest here among the wetland meadows but others will rest, feed up and carry on further north, across the border of Russia and on to the Siberian Arctic tundra.

An Dao was able to show his son three such species stopping over in Daxing'anling—the bean goose and their colleagues the Whooper and smaller Bewick swans.

豆雁是一只体型很大的雁，嘴尖有一明显黄斑。它们要去北方繁殖，但到大兴安岭时早已成双结对了，在这里降落后它们又双双对对聚集成群。

豆雁群一起落下，在农田或湿地上进食和休息。有时它们就落脚在湖泊和河流上正融化的冰块上。它们的脚皮厚，感觉不到冷。也许在水中它们可以远离那些掠食动物——狐狸、狼和獾，它们会相对安全。

巴拉的父亲解释了这些不同种类鸟群之间的区别。

小天鹅飞行旅程最长。它们会一直飞到西伯利亚北极，在那里分散开来，在苔原上的稀疏植被中筑巢。

小天鹅一身白色，很漂亮，嘴黑色，嘴基有黄色点缀。它们也成双配对，但可能还有去年的幼鸟跟随在身边，幼鸟的头还有些黄色。小天鹅跟配偶可能不离不弃在一起多年。

但最壮观的要算大天鹅。它们看起来跟小天鹅相似，但是它们体型较大，颈部很直。大天鹅的嘴更长，嘴基黄色呈长三角形，一直延伸到黑色的鼻孔处。

大天鹅不像小天鹅那样飞那么北，有些大天鹅对儿可能就在大兴安岭的沼泽和湿地上找到合适的地方筑巢。它们会收集芦苇和草，堆积成一大堆，然后产8~12枚卵。

The bean goose is a big goose with a conspicuous yellow mark at the tip of its bill. By the time they arrive at Daxing'anling on their way to breed, they are already mostly in pairs, but pairs gather in larger flocks.

The bean goose flocks drop down to feed and rest on the farmlands or wetlands. Sometimes they will simply rest on the sheets of ice still melting on the lakes and bigger rivers. They seem to feel no cold in their leathery feet and they feel safer from possible preadators—foxes, wolves, wolverines above the water.

Ba La's father explained some of the differences between the different bird guests.

The Bewick swans have the longest journey. They will fly on right up to the Siberian Arctic where they spread out and nest in the sparse vegetation of the tundra wastes.

Bewicks are attractive white swans—all white but with a black bill decorated by a yellow base. They two are paired up but may be accompanied by last year's youngsters who still have some orange tinge to their head. Pairs may remain together for many years.

But the most magnificent are the Whooper swans. They look similar to the Bewicks but they are larger with very straight erect necks. The bill of the whooper is longer and the yellow base forms a long triangle that extends even more forward that the black nostril.

The Whoopers do not fly as far north as the Bewicks and indeed some pairs may find suitable nesting sites among the bogs and wetland marshes of Daxing'anling. Here they will collect and heap up a substantial pile of reeds and plucked grasses and then lay 8-12 eggs.

大天鹅　　小天鹅

　　到夏天如果你再来看，就能看到它们在自豪地领着身后一群毛茸茸的灰色小天鹅游泳。

　　夏季白天长，水中杂草和嫩草生长迅速，食物丰富，雏鸟成长也非常快。它们也吃一些在沼泽里很容易抓到的蚊、蠓和水苍蝇等昆虫。

　　秋天，大兴安岭又准备好了迎接在北方繁育、现在要南迁的鸟群。只要还有食物，水还没冻结，鸟群就会在这里停留。

　　豆雁喜欢吃秋收后田间留下来的任何谷类或豆类。而天鹅吃得更多的是绿色的叶子。

If you come back in the summer you can see the whooper swans swimming proudly with their broods of fluffy grey coloured cygnets.

With long days in summer, fast growth of water weeds and tender herbs, there is plenty of food and the young grow quickly. They add a few insects easily caught among the great swarms of mosquitoes, midges and water flies.

By autumn the Daxing'anling bird hotel is ready to receive the returning northern breeders gathering in big flocks ready for their journey south. The flocks will stay as long as there is still food and there is still open water.

The bean geese will eat any grain or beans left behind in the fields after the harvest. The swans eat more green leaves.

冬天的第一次霜冻到了，湖泊和一些慢流的小溪流就会开始冻结。永久冻土下的冰层开始膨胀，植物停止生长。这时鸟类就继续南行了。

大群的水禽拍打着翅膀，积聚力量，呼呼呼呼地开始了南行之旅。大部分的鸟要再飞2000千米，到长江流域找到不结冰的湖泊，在那里越冬。

"鹤类呢？"小巴拉问道："为什么我们这里很少能看到鹤呢"。

"鹤类啊，"安岛说："确实中国东北地区对鹤类也很合适，但是它们更喜欢在潮湿的草地上生活，它们并不怎么喜欢林地。"

在这里的田野上偶尔可以看到白枕鹤和灰鹤，但你在呼伦贝尔可以看到白头鹤和优雅的蓑羽鹤的鸟巢。再往南，漂亮的丹顶鹤会在黑龙江的扎龙地区筑巢。

"我们能去那里看看吗？"小巴拉满怀希望地问。

爸爸看着那个小男孩。他很高兴巴拉对野生动物非常感兴趣。

"好的，等你学校一放假，我就带你去呼伦贝尔看鹤。"

小巴拉一听，高兴极了。

With the first frosts of winter the lakes and slower streams start to freeze over. The underlying ice of the permafrost wetlands starts to expand once again and plant growth stops. It is time to move on.

With much flapping and honking the flocks of great waterfowl summon up their energy for the journey south. It is another 2000km till they can find unfrozen lakes in the Changjiang Valley where most of them will spend the winter.

"What about cranes?" asked Xiao Ba La. "Why do we not see many cranes here?"

"Ah!" said An Dao. "Indeed NE China is a great region for cranes but they prefer the wet grasslands and not so much the forest."

We can see the occasional White naped crane and common cranes in the fields here but you need to go up to Hulunbier to see the nesting areas of Hooded crane and the pretty Demoiselle cranes. Or south of here to Zhalong in Heilongjiang to see the nesting area of the magnificent Red-crowned crane.

"Can we go and see them?" asked Ba La hopefully.

An Dao looked at the little boy. He was so happy that Ba La was taking so much interest in wildlife.

"Ok. we wait for your school holiday and then I will take you to see the cranes in Hulunbuir."

Ba La looked very happy at the thought.

小测试

1. 找出大天鹅和小天鹅之间的三个不同点。

2. 这两种天鹅中哪一种在大兴安岭繁殖?

3. 天鹅的雏鸟是什么颜色的?

4. 天鹅在树上筑巢吗?

5. 豆雁吃什么食物?

6. 为什么水禽在冰层上落脚?

7. 鹤到哪儿去繁殖?

8. 迁徙的水禽去哪儿越冬?

9. 什么时候是观看迁徙水禽的最佳时候?

答案

1. 大天鹅体型比较大,嘴基黄色较多,不飞到北极;

2. 大天鹅;

3. 灰色,羽毛蓬松;

4. 不,它们的巢就是在近水处的地面上大堆的植被;

5. 豆类、谷物和蔬菜;

6. 在食肉动物不敢去捕捉它们的地方,它们觉得更安全;

7. 开阔的湿地,如扎龙和呼伦贝尔;

8. 多数鸟飞到长江河谷;

9. 秋天和春天。

Quizz Time

1. List three differences between the greater and lesser swans.

2. Which of these swans breeds in Daxing'anling?

3. What colour are young swans?

4. Do swans nest in trees?

5. What do the Geese eat?

6. Why do the waterbirds rest on ice sheets?

7. Where do the cranes go to breed?

8. Where do most migrating waterfowl go for the winter?

9. When is the best time to see migrating waterbirds?

Answers

1. The whooper is larger, has more yellow on its bill and does not fly all the way to the Arctic;

2. The larger Whooper swan;

3. Grey and fluffy;

4. No, they make big nests of vegetation on the ground near water;

5. Beans, grain, and vegetables ;

6. They feel safer where carnivores dare not chase them;

7. Open wetlands such as Zhalong and Hulubier;

8. Most go to the lakes in the Changjiang Valley;

9. Autumn and spring.

紫闪蛱蝶奇特的生命周期

大兴安岭的蝴蝶种类特别丰富。因为夏天极其短暂，所有蝴蝶几乎在一年的同一时间出来，令人眼花缭乱。

此时各个物种都忙忙碌碌寻找合适的配偶，交配后雌蝶还必须找到有食物的植物去产卵。

孩子们常听说蝴蝶的生命只有一天。其实这不对。几乎所有蝴蝶都能活一个月以上，有些蝴蝶可以活好几个月。

巴拉在小河边玩儿。他朝他的叔叔喊："毛歇叔叔，快来看这些好大的蝴蝶。三只蝴蝶在一块儿喝水，它们真美。"

毛歇过来一看，果然是三只燕尾蝶在小河边潮湿的岩石上吸吮水分。

"你想它们为什么都在同一处吸水？"毛歇叔叔问道。"可能是它们相互喜欢吗？"巴拉猜想。

The Curious Life Cycle of the Large Blue Butterfly

Butterflies are abundant and often dazzling in Daxing'anling and as summer is short, all the butterflies come out at about the same time of year.

Each species has a busy time to find an appropriate mate, then the female has to find the right food plant on which to lay her eggs.

Children are often told a butterfly only lives for a day. But this is not true. Some butterflies can live for several months and almost all live for at least one month or more.

Ba La is playing by the stream. He calls his uncle.

"Uncle Mao Xie come and see these big butterflies. There are three drinkng together. They are so beautiful."

Mao Xie came over to see. They were indeed three Swallowtails sipping water from the wet rocks by the stream side.

"Why do you think they all drink at the same place?" Uncle Mao Xie asked.

"Maybe they like each other?" guessed Ba La.

"不，它们并不是合群的种类。它们在这里找到了土壤里渗出的一些特殊的矿物质。如果它们能不断补充它们所需的矿物质，它们就可以活得更长。"毛歇叔叔说。

蝴蝶的生活看似很简单，只是飞来飞去，找个配偶而已，但其实它们的生活复杂多了。如果天气炎热，蝴蝶可能变干，那么它可能只能活一天。因此蝴蝶一直在找水喝，最好是富含矿物质的水来啜饮和补充它们的膜组织。下雨后，在路边的溪流你可以看到有大群的蝴蝶聚集一起。它们会把啄管伸出很长以吸取液体，多余的水分则从肛门喷出，这样它们就能从所吸取的液体中留下少量的营养物质。

"你仔细看这三只燕尾蝶，每10秒左右就有水滴从尾部喷出。"

其他蝴蝶种类通过从花卉上吸取富含糖分的花蜜来延长它们的生命。有些蝴蝶有异常的喜好。通常你会看到蝴蝶聚集在麝猫或其他哺乳动物的粪便上吸取液体。美丽的闪蛱蝶甚至喜欢吸食动物死尸的液体。

巴拉和他的叔叔观看着各式各样的蝴蝶，毛歇叔叔一边详细地教巴拉蝴蝶的生命历程。他们的这个下午过得很开心。

不同种类的蝴蝶已经演化成能在不同的植物上取食。有些蝴蝶在森林中的柳树或栎树上吸食。有些蝴蝶喜欢草地上的草和药草。不同的植物含有不同的化学物质，其中有些是有毒的。不同的蝴蝶物种由于进化，已经具有对一些化学物质的解毒能力。在某些情况下，它们会将毒素吸收到自己的身体内，然后加剧毒力。这种毒素会给它们以保护，因为有些以昆虫为食的鸟类，学会了避免吃这些难吃的昆虫。

"No they are not very sociable. They have found a place where they can get some special minerals seeping from the land in this place. They can live longer so long as they can keep topping up their supply." said Mao Xie.

You may think their life is simple. Just fly about and look for a partner but life can be a bit more complicated than that. On a hot dry day a butterfly might indeed survive only one day if it dries out. So the butterflies look for water and even mineral rich water to sip and replenish their tissues. You can see great flocks of butterflies gathering at streams along the roads after rain. They extend their long proboscis to suck up the liquids and may squirt out the excess water from their anus as they collect rare nutrients from their intake.

"Take a closer look at our three swallowtails. They are squirting droplets of water from their backside every 10 seconds or so."

"Other species prolong their active lives by sucking sugar-rich nectar from sweet flowers. Some other species have even more unusual tastes. Often you will see butterflies gathering to drink from the dung of civets or other mammals. The rather beautiful Purple Emperor butterfly is even rather fond of feeding on the corpses of dead animals."

Ba La and his uncle had a lovely afternoon looking at different butterflies whilst uncle Mao Xie tried to teach Ba La some of the details of their life histories.

Different butterfly species have evolved to feed on different plant species. Some select willows or oaks in the forests. Others prefer grasses and herbs in the meadows. Different plants have different chemicals and some of these are poisonous. Different butterfly species have evolved ability to tackle these poisonous chemicals. In some cases they absorb the chemicals into their own bodies and become more poisonous themselves. This gives them some protection from insect-eating birds that learn to avoid the more unpalatable insects.

蝴蝶的生命周期经历四个阶段：卵—幼虫—蛹—成虫蝴蝶。

蝴蝶的卵需要几天的时间才能孵化，成为微小的幼虫或毛毛虫。它们开始会吃卵壳，然后开始吃寄主植物的软组织。

蝴蝶幼虫阶段看起来根本不像成虫蝴蝶。随着幼虫长得越来越大，它的皮肤变得越来越紧，最终开裂，软软的蛹出现了，一层更大的蛹皮又逐渐变硬。饥饿的毛毛虫继续吸食，继续长大。

大多数蝴蝶都要经过6~7次这样虫和蛹的蜕变，最终才变成蝴蝶。

蛹就像装在盒子里的变形机器人。在完成变态之前，蛹可能会待在盒子里几周，在冬天甚至会冬眠。

但蛹里面的结构却在发生巨大的变化。蛹体内的体液和毛虫的结构会完全重新组织，复杂的蝴蝶成虫结构开始形成——头、胸、腹、翅、腿和触须。最后开始形成彩色的鳞片，通过蛹的透明外壳可以看到蝴蝶翅膀的图案。

There are four stages in a butterfly's life cycle: eag—larvae—capterpillar—butterfly.

Butterfly's eggs take a few days to hatch and then the tiny larvae or caterpillars emerge and start eating first the egg shells and then the soft tissues of the host plant.

The butterfly larval stage looks nothing like the adult butterfly. As the caterpillar grows larger, its skin gets tighter and tighter until the skin eventually splits, the soft larva emerges and a new larger sized skin hardens and the caterpillar carries on feeding and growing.

Most butterfly species do this about 6 or 7 times before they change form completely and pupate.

The pupa is like a transformer robot in a case. The pupa may remain for several weeks or may even hibernate through the winter before completing its development.

Great changes take place inside the pupa case. The body fluids and contents of the caterpillar are becoming completely re-organised and the complex structures of the adult butterfly start to form - head, thorax, abdomen, wings, legs and antennae. Eventually even the coloured scales start to form and the pattern of the butterfly wing can be seen through the transparent shell of the pupa.

蝴蝶生命周期

最后，季节到了，蛹的外壳会分裂，成虫蝴蝶就出来了。蝴蝶的翅膀刚开始又小又软，无法飞翔。但蝴蝶通过翅膀上的静脉输送液体，翅膀就像一辆旧自行车的轮胎一样膨胀硬朗起来。

蝴蝶要等翅膀充分展开、干燥、变硬之后，才能飞行。蝴蝶会先试飞几次，才能展翅在空中开始飞行。

甲虫和苍蝇也要经过类似的过程，这个相当复杂的生命周期叫作"完全变态"。有的昆虫，比如蟑螂、螳螂和蜻蜓，只做不完全的变态，因为它们的宝宝看起来像小成虫，每次从卵鞘里冒出，虫体变得越来越大，直到成虫翅膀长成。但所有蝴蝶都要经历从卵到毛虫到蛹直到成虫蝴蝶的整个周期——完全变态周期。

过了两天，毛歇叔叔叫上巴拉："巴拉跟我来，我带你去看看一些神奇的东西。"

毛歇领着巴拉上了山，来到一片长满草的开阔坡地，在这里可以俯瞰下面的山谷。这里有无数美丽的花卉，上面有数不胜数的蝴蝶和蜜蜂在飞舞。但是叔叔没有关注这些，而是带着巴拉来到了一片开着粉色碎花的香草前。他采起一小枝，用手指捻了捻，递过去给巴拉闻。

"你知道这种香草吗？这叫百里香。你奶奶用它来炒菜，也用它来做药。"

这个香味巴拉熟悉，但他以前并不认识这种野花植物。一只漂亮的阿波罗绢蝶——一种很大的、身上有美丽的黑红两色眼斑的白色蝴蝶，正在这些百里香的花上取食。但是毛歇叔叔要巴拉看的是一种小型的紫闪蛱蝶。这种蛱蝶双翼展开只有4厘米宽，它熠熠发光，翼上有黑色的粗点斑。

Finally at the right season the case will split, and the adult butterfly emerges. At first the wings are small and floppy and the butterfly is unable to fly. But the insect pumps fluid through the veins of its wings and they inflate a bit like the tyres of an old pump bicycle.

The butterfly has to allow the fully expanded wings to dry off before they are rigid enough to fly. The insect gives a few tentative flaps and then launches off into the air for its first aerial flight.

This rather complicated life cycle is called a 'complete metamorphosis'. Beetles and flies go through a similar transformation, but some other insects such as cockroaches, grasshoppers and dragonflies, only do an 'incomplete metamorphosis' in that their babies hatch as nymphs that look like tiny adults and each emergence from its skin just allows the insect to get bigger and bigger till the last stage adult develops wings. But butterflies all undertake the complete change from egg to caterpillar to pupa and finally adult.

Two days later uncle Mao Xie called Aoli. "Come with me Ba La. I will show you something really amazing."

Mao Xie led the way up the hill towards the open grassy slopes overlooking the valley. There were so many pretty flowers and so many butterflies and bees feeding on them. But uncle ignored all these until he found a patch of small pink flowering herbs. He picked a shoot, squeezed it in his fingers and held it out for Ba La to smell.

"You know that plant. It is thyme and your grandma uses it in her cooking and to make medicines."

Ba La recognized the smell but had never recognized the plant as a wild flower. A pretty Apollo butterfly-big, white and decorated with beautiful black and red eyespots was feeding on the thyme flowers but it was the little blue butterfly that uncle wanted to show Ba La. It was a small fellow just 4 cm across; a shimmering blue with bold black spots on its wings.

"这只小蝴蝶却把'完全变态'过程更推进一步，使其更加不可思议。"他解释道。

紫闪蛱蝶成虫蝴蝶的行为跟普通蝴蝶一样，它会交配，也从鲜花中吸吮花蜜来补充能量。雌蝶把卵产在她最喜爱的寄主植物上，比如可以用来炒菜的百里香。

紫闪蛱蝶的幼虫期也算正常，但是它们宁吃百里香的花，而不吃嫩叶。到了第三期（蛹）的时候，它的蜕变就变得很奇怪。

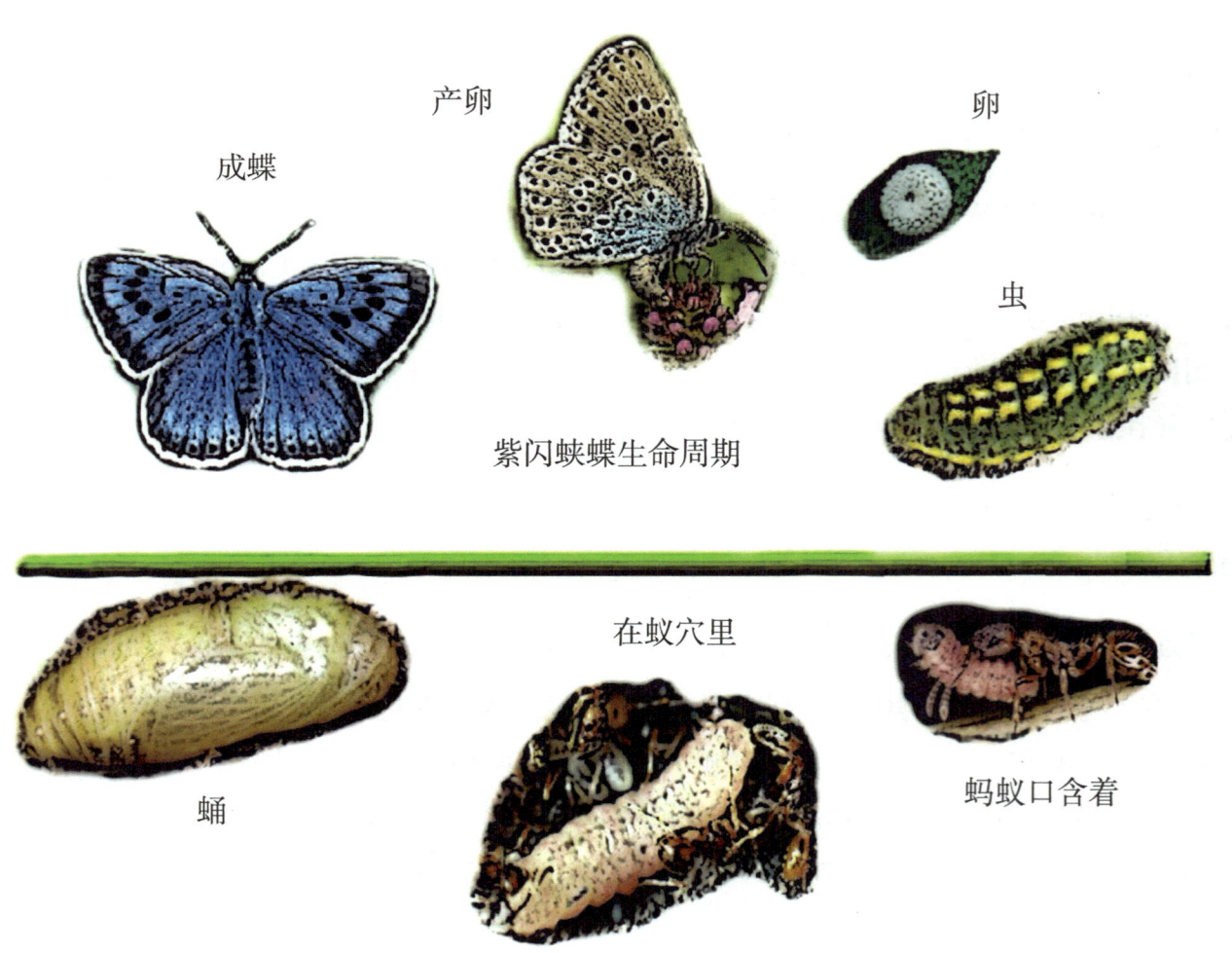

紫闪蛱蝶生命周期

"This little butterfly takes the idea of complete metamorphosis a whole step further." he explained.

"The adult behaves like a normal butterfly; it mates, replenishes its energy by sucking nectar from flowers and the female lays her eggs on her favourite food plant—the kitchen herb wind thyme.

All remains normal for the young caterpillars although they eat flowers and even seeds of the thyme plants in preference to the delicate leaves. After the third stage things go a bit bizarre."

叔叔接着给巴拉解释。紫闪蛱蝶的幼虫从寄主植物上掉下来，等待着路过的红蚂蚁把它捡起，然后运送到蚁穴中。这时蝴蝶幼虫已经演变成貌似红蚂蚁的蛹，甚至发出跟蚂蚁一样的气味。蚂蚁的反应呢，似乎是它发现了一只丢失了的蚂蚁幼虫，因此就把这幼虫带回，放到自己的幼虫群中。一入蚁穴，紫闪蛱蝶的幼虫就会找到巢室，开始吃红蚂蚁的卵和幼虫。冬天来临，这只大的紫闪蛱蝶幼虫就开始休眠。

春天复苏时，紫闪蛱蝶继续在蚁穴中吃红蚂蚁卵和幼虫，直到它准备变成蛹。最后当天气变暖时，紫闪蛱蝶成虫就会涌现，整个奇特的生命周期又周而复始。

这还不止。紫闪蛱蝶幼虫吃蚂蚁有两种不同的策略。有些紫闪蛱蝶只吃对它们很慷慨的蚂蚁的卵和幼虫，但有些紫闪蛱蝶更不像话，它们会模仿蚁后发出的声音。这使它们具有显要身份，工蚁们就直接用糖和其他食物来供养这些紫闪蛱蝶幼虫。

自然演变中真有很多不可思议的东西！

回家的路上巴拉一直沉默不语，他需要好好整理一下自己思绪。

Uncle then described to Ba La how the caterpillar drops off the food plant and waits to be picked up by a passing red ant, which will then carry it to its nest. The butterfly larva has evolved to look like the pupa of the red ant and even emits a smell that copies the ant smell. The ants react as though they found a lost ant larva and take the caterpillar back to their own larval nursery. Soon after it enters the nest, the caterpillar will seek out the nesting chambers and proceed to eat red-ant eggs and larvae. When winter draws near, the large blue caterpillar begins to hibernate.

Next spring it carries on feeding in the ant nest until it is ready to emerge and pupate. Finally when the weather gets warm enough the adult butterfly will emerge and the whole curious life cycle starts again.

Even this is not the entire story. Large blue caterpillars have two different strategies for feeding in the ant colony. Some individuals simply eat the eggs and larvae of their generous host ants but other individuals are even more ungracious. They make a sound like the sound of the larva of the queen ant. This gives them VIP status and worker ants then come to feed the growing caterpillar directly with sugars and other foods.

Nature truly has evolved some amazing ways of doing things!!

Ba La was silent as they walked home. He had a lot of thoughts to sort out.

小测试

1. "完全变态"的意思是什么？

2. 你能说出另一种"完全变态"的昆虫吗？

3. 你能说出一种"不完全变态"的昆虫吗？

4. 为什么有些蝴蝶叮在森林里动物的粪便上？

5. 燕尾蝶怎么越冬？

6. 为什么蚂蚁会收集和照顾紫闪蛱蝶的幼虫？

7. 有些蝴蝶如何变得对鸟类来说很难吃？

答案

1. 完全变态的意思是动物从卵变成幼虫，幼虫变成蛹，最终变成成年动物的形态；

2. 甲虫和苍蝇有完全变态的过程；

3. 蜻蜓、螳螂和蟑螂都有不全变态，它们有若虫阶段，没有幼虫和蛹的变态；

4. 它们用啄管吸取重要的营养物质，以延长它们的生命；

5. 以蛹的形态，挂在寄主植物的枝干上；

6. 幼虫看上去、闻起来都像是蚂蚁的幼虫，有些蝴蝶幼虫甚至能分泌出蚂蚁喜欢舔食的甜味奶汁；

7. 它们从寄主植物上吸取有毒化学品，保留在体内。

Quizz Time

1. What is the meaning of 'complete metamorphosis'?

2. Name another insect that has 'complete metamorphosis'.

3. Name an insect that has 'incomplete metamorphosis'.

4. Why do some butterflies sit on the dung of forest animals?

5. How does the swallowtail butterfly pass the winter?

6. Why do ants collect and care for butterfly larvae?

7. How do some butterflies become distasteful to birds?

Answers

1. Complete metamorphosis means the animal transforms itself from an egg to a larva and then a larva to a pupa and finally emerges as an adult;

2. Beetles and flies also have complete metamorphosis;

3. Dragonflies, mantids and roaches all have incomplete metamorphosis having nymphal stages instead of larva and pupa;

4. They collect vital nutrients that they can inject and helps prolong their life;

5. As a pupa, hanging from the stem of its foodplant;

6. The larva looks like and smells like an ant grub, and some larvae even exude a sweet milk that the ants can enjoy licking;

7. They absorb toxic chemicals from their foodplants and hold them in their own tissues.

如何保护大兴安岭美丽的野生动物

来吧，让我们进入大兴安岭的森林，亲自去发现大自然的奇迹，亲身去感受生活在这里的神奇生物。

你可以沿着更安静的道路行驶，或者更好地沿着小道和小径行走，真正体验到这里的自由和空间。闻花香，听鸟语，看蝴蝶飞舞，忘记自己，让自己融入这个绝妙的野生世界里。

要确保你的安全。穿适宜的服饰，可能会突然下雨或变冷，要带上雨衣或雨伞。穿适宜的靴子。

带上一个朋友或让别人知道你要去的地方。提前了解你的行程和路线。带上指南针，也可以把手机作为指南针和地图用，但别忘了，大兴安岭的边远地区还没有手机信号。

夏天，特别是在清早和傍晚，有很多会咬人的苍蝇和蚊虫。你可以戴上有面纱的帽子来保护脸部，或者在出门前喷洒驱蚊剂。

每隔几分钟要查看一下裤子，看看有没有草爬子爬上了你的腿。如果你被草爬子咬了，你应该在医务人员的帮助下把它们拔除，如果你自己随便把它弄出来，它们的嘴巴的残余部分可能会留在你的皮肤里，这会导致皮肤腐烂，引起感染。

带上望远镜，这样你就可以更好地欣赏鸟类和动物。

带上笔记本，这样方便记录你看到的东西。如果有什么发现，把它画下来，多年以后你可能还会觉得这些记录很有意思。随着气候的变化，很多季节性发生的事情会变得越来越早。

享受大自然最重要的规则是不要打扰或破坏自然。保持安静。其他游客可能喜欢寂静和自然的声音。他们可不想听到你大喊大叫！不要太靠近鸟类和动物，也不要采摘花卉。

除了在规定的野营区，不要点火燃火。垃圾要自己带走或放到垃圾容器里。

如果要乘船或漂流，务必穿上防护背心。按照主管人的指示做就是。但最重要的是好好享受好好玩。

如果你学会了欣赏自然，我们希望你会的。你还可以参与一些社团和保护工作，充分利用你能提供的帮助，发挥你的精力。你可以写写你的体验。你可以跟你的朋友说说，争取更多的帮助，更好地保护自然。

拒绝吃野生动物或受保护的物种。不在任何水道倒垃圾或有毒物质。

热爱生命，保护自然。

How You Can Help Protect the Beautiful Wildlife of Daxing'anling

So come into the woods and discover for yourself the wonders of nature and the fascinating creatures that live in Daxing'anling.

Drive along the quieter roads or better still walk along the lanes and trails to really experience the freedom and space. Smell the flowers, hear the birds, watch the butterflies, forget yourself and become part of this great wide world.

Make sure you are safe. Wear sensible clothes. It can suddenly rain or get cold. Have a rain cape or umbrella. Wear good boots.

Take a friend or let people know where you are going to travel. Have a good idea of your route in advance. Take a compass. You can use a cellphone as compass and map but remember the remoter parts of Daxing'anling still have no signal.

In summer there are biting horseflies and mosquitos especially in the early morning and evening. You can wear a hat which has netting to protect your face or you can spray yourself with insect repellent before you set out.

Look down at your trousers every few minutes. There may be ticks climbing up your legs. If they bite you seek help in removing them because if you pull them roughly they may leave part of their mouthparts in your skin and this can decay and become infected.

If you have binoculars you can appreciate the birds and animals better.

Keep a notebook so you can record and remember what you see. Make little drawings of the things you find. Even years later you will find these old notes interesting. Seasonal events will get earlier with climate change as years pass.

The main rule of enjoying nature is to not disturb or destroy nature. Keep quiet. Other visitors may be enjoying the peace and natural sounds. They do not want to hear you shouting! Do not get too close to birds and animals. Do not break or collect flowers.

Do not light any fires except in prescribed camping areas. Carry out your own litter or use litter receptacles provided.

If you go on a boat or raft, make sure to wear proper safety vest. Do what your supervisor tells you. But most of all enjoy yourself.

If you learn to love nature as we hope you will, there are societies to join and conservation works that can absorb your help and energy. Write about your experience. Tell your friends. Get more help to protect nature better.

Do not eat wildlife or protected species. Do not tip rubbish or toxic materials into any waterways.

Love Life, protect your nature.